# The Practical Beekeeper

## Volume II
## Intermediate
## Beekeeping Naturally

### by Michael Bush

The Practical Beekeeper Volume II
Intermediate Beekeeping Naturally

Copyright © 2004-2011 by Michael Bush

All rights reserved. No part of this book may be reproduced or transmitted in any form or by any means without written permission of the author.

Cover Photo © 2011 Alex Wild www.alexanderwild.com

ISBN: 978-161476-062-7

# X-Star Publishing Company
Nehawka, Nebraska, USA
xstarpublishing.com

218 pages

# Dedication

This book is dedicated to Ed and Dee Lusby who were the real pioneers of modern natural beekeeping methods that could succeed with the Varroa mites and all the other new issues. Thank you for sharing it with the rest of us.

# About the Book

This book is about how to keep bees in a natural and practical system where they do not require treatments for pests and diseases and only minimal interventions. It is also about simple practical beekeeping. It is about reducing your work. It is not a main-stream beekeeping book. Many of the concepts are contrary to "conventional" beekeeping. The techniques presented here are streamlined through decades of experimentation, adjustments and simplification. The content was written and then refined from responding to questions on bee forums over the years so it is tailored to the questions that beekeepers, new and experienced, have.

In place of an index there is a very detailed Table of Contents.

It is divided into three volumes and this edition contains only one: Volume II, Intermediate Beekeeping Naturally.

# Acknowledgments

I'm sure I will forget to list many who have helped me along this path. For one thing many were often only known by the names they used on the many bee forums where they shared their experiences. But among those who are still helping, Dee, of course, Dean and Ramona, and all of the wonderful people of the Organic Beekeeping Group on Yahoo. Sam, you are always an inspiration. Toni, Christie thanks for your encouragement. All of you on the forums who asked the same questions over and over, because you showed me what needed to be in this book and motivated me to write down the answers. And of course all of you who insisted I put this in the form of a book.

# Foreword

I feel like G.M. Doolittle when he said he had already offered all of what he had to say for free in the bee journals and yet people kept asking for a book. I have virtually all of this on my website and have posted all of it many times on the bee forums. But many people have asked for a book. There is a little new here, and most of it is available for free already on my web site (www.bushfarms.com/bees.htm). But many of us understand the transient nature of the medium of the web and want a solid book on our shelf. I feel the same. So here is the book that you could already have read for free but you can hold it in your hands and put it on the shelf and know you have it.

I've done a lot of presentations and a few have been posted on the web. If you have an interest in hearing some of this presented by me try a web search for videos for "Michael Bush beekeeping" or other topics such as "queen rearing". The material here is also on www.bushfarms.com/bees.htm along with PowerPoint presentations from my speaking engagements.

# Table of Contents

**Volume II Intermediate** ............ **259**

**A System of Beekeeping** ............ **261**
- Context ............ 261
- Examples ............ 262
- Locale ............ 262
- Why a system? ............ 263
- Integration and related issues ............ 263
- Why this system? ............ 263
- Sustainable ............ 263
- Workable ............ 264
- Efficient ............ 264

**Decisions, Decisions...** ............ **265**
- Kinds of beekeeping ............ 265
  - Commercial ............ 265
  - Migratory ............ 265
  - Fixed ............ 265
  - Sideliner ............ 266
  - Hobbyist ............ 266
- Personal Beekeeping Philosophy ............ 266
  - Organic ............ 266
  - Chemical ............ 267
  - Science vs. Art ............ 267
  - Scale ............ 268
  - Reasons for beekeeping ............ 268

**Locality** ............ **269**
- All Beekeeping is Local ............ 269

**Lazy Beekeeping** ............ **271**
- Top Entrances ............ 272
- Uniform frame size ............ 274
- Lighter boxes ............ 276
- Horizontal hives ............ 280
- Top Bar Hive ............ 281
- Foundationless frames ............ 282

Making foundationless frames .................................282
  No chemicals/no artificial feed....................................284
  Leave honey for winter food ......................................285
  Natural cell size............................................................286
  Carts ............................................................................287
  Leave the burr comb between boxes.........................287
  Stop cutting out swarm cells ......................................289
  Stop fighting your bees ..............................................290
  Stop wrapping your hive. ...........................................290
  Stop scraping all the propolis off of everything............291
  Stop painting your equipment. ...................................292
  Stop switching hive bodies. ........................................293
  Don't look for the queen.............................................294
  Don't wait. ...................................................................295
  Feed dry sugar............................................................296
  Split by the box...........................................................296
  Stop Requeening.........................................................297

**Feeding Bees ........................................................ 298**
  First, when do you feed? ............................................298
  Stimulative feeding.....................................................300
  My experiences with stimulative feeding....................302
    Down sides to success: .........................................303
    Variable outcomes:................................................304
    Dry Sugar: .............................................................304
    Type of feeder: ......................................................305
  Second, what do you feed?........................................305
    Pollen ....................................................................306
  Third, how much do you feed? ...................................306
  Fourth, how do you feed?...........................................306
  Issues when considering the type of feeder: ..............307
  Basic types of feeders ................................................308
    Frame feeders........................................................308
    Boardman feeder...................................................309
    Jar feeder...............................................................309
    Miller feeder ..........................................................311
    Bottom board feeder..............................................313
    Jay Smith Bottom Board Feeder ............................313
    My version.............................................................315

Baggie feeder ........................................................... 320
Open feeder ............................................................ 321
Candy board ........................................................... 321
Fondant .................................................................. 321
Dry Sugar ............................................................... 321
What kind of sugar? ............................................... 326
Pollen ..................................................................... 327
Measuring ratios for syrup .................................... 327
Weight or Volume? ................................................ 327
How to measure .................................................... 328
How to make syrup ............................................... 328
Moldy syrup .......................................................... 329

# Top Entrances ............................................... 330
Reasons for top entrances .................................... 330
How to make top entrances .................................. 332
Top Entrance Frequently Asked Questions: ........ 335

# Carts ............................................................... 338

# Swarm Control ............................................. 342
Causes of swarming ............................................. 344
Overcrowding swarm ............................................ 344
Reproductive swarm ............................................. 345
Preventing swarming ............................................ 347
Opening the broodnest ......................................... 347
Checkerboarding aka Nectar Management ......... 348

# Splits .............................................................. 349
What is the desired outcome? .............................. 349
Timing for doing a split: ........................................ 349
The concepts of splits are: ................................... 350
Kinds of splits ....................................................... 351
An even split ......................................................... 351
A walk away split .................................................. 351
Swarm control split .............................................. 351
A cut down split .................................................... 352
Concepts of a cut down: ...................................... 352
Confining the queen ............................................. 353
Cut down Split/Combine ...................................... 353

Frequently Asked Questions about splits............354
  How early can I do a split?............354
  How many times can I split?............354
  How late can I do a split?............355
  How far?............356

# Natural Cell Size ............ 357
And its implications to beekeeping and Varroa mites.....357
Does Small Cell = Natural Cell?............357
  Baudoux 1893............357
  Eric Sevareid's Law............357
  Foundation Today............358
Chart of Cell Sizes............362
Volume of cells............363
Things that affect cell size............364
What is Regression?............364
How do I regress them?............364
Observations on Natural Cell Size............366
Observations on Natural Frame Spacing............367
  $1\frac{1}{4}$" spacing agrees with Huber's observations............367
  Comb Width (thickness) by Cell Size............367
Pre and Post Capping Times and Varroa............369
Huber's Observations............370
My Observations............370
Why would I want natural sized cells?............371
How to get natural sized cells............371
How to get small cells............371
So what Are natural sized cells?............372
Conclusions:............372
Frequently asked questions:............372
So let's do the math:............376

# Ways to get smaller cells............ 380
How to get natural sized cells............380
  Top Bar Hives............380
  Foundationless frames............380
  Make blank starter strips............380
How to get small cells............380

# Rationalizations on Small Cell Success ...... 381

AHB ........................................................................ 381
Survivor stock ........................................................ 382
Blind faith ............................................................... 383
Resistance .............................................................. 384
Small Cell Studies .................................................. 385

**Foundationless** ............................................... **386**
Why would one want to go foundationless? ................. 386
How do you go foundationless? ................................... 387
Foundationless frames ................................................ 391
Historic References ..................................................... 391
FAQs .............................................................................. 395
    Box of empty frames? ............................................. 395
    What is a guide? ....................................................... 395
    Best guide? ................................................................ 395
    Extract? ...................................................................... 396
    Wire? .......................................................................... 396
    Wax them? ................................................................ 397
    Whole box? ............................................................... 397
    Will they mess up? ................................................... 397
    Slower? ...................................................................... 398
    Beginners .................................................................. 399
    If they mess up? ....................................................... 399
    Dimensions ............................................................... 400

**Narrow Frames** ............................................... **401**
Observations on Natural Frame Spacing ..................... 401
    1 $\frac{1}{4}$" spacing agrees with Huber's observations ........ 401
    Comb Width by Cell Size ......................................... 402
Historic references to narrower frame spacing ............. 402
Spacing frames 1$\frac{1}{4}$" has advantages ........................... 407
Frequent misconceptions: ............................................. 408
Ways to get narrow frames ........................................... 408
FAQs .................................................................................. 409

**Yearly Cycles** .................................................... **411**
Winter ............................................................................. 411
    Bees ............................................................................ 411
    Stores ......................................................................... 412
    Setup for winter ....................................................... 412

Spring..................................................................................412
  Summer.............................................................................413
  Fall......................................................................................413

# Wintering Bees ........................................................ 415
  Mouse Guards...................................................................415
  Queen Excluders ..............................................................416
  Screened bottom boards (SBB) ......................................416
  Wrapping ..........................................................................417
  Clustering hives together................................................417
  Feeding Bees ...................................................................417
  Insulation .........................................................................419
  Top Entrances..................................................................420
  Where the cluster is ........................................................420
  How strong? ....................................................................420
  Entrance reducers ...........................................................421
  Pollen ................................................................................421
  Windbreak........................................................................422
  Eight frame boxes ...........................................................422
  Medium boxes.................................................................422
  Narrow frames ................................................................423
  Wintering Nucs................................................................423
  Banking queens ...............................................................424
  Indoor wintering..............................................................425
  Wintering observation hives..........................................425

# Spring Management................................................ 426
  Tied to climate .................................................................426
  Feeding Bees ....................................................................426
  Swarm Control.................................................................427
  Splits ..................................................................................428
  Supering ...........................................................................429

# Laying Workers ....................................................... 430
  Cause .................................................................................430
  Symptoms.........................................................................430
  Solutions ...........................................................................431
    Simplest, least trips to the beeyard ........................431
    Shakeout and forget ..................................................431
    Most successful but more trips to the beeyard..........431

    Give them open brood ................................................. 431
    Other less successful or more tedious methods ......... 432
  More info on laying workers ........................................... 433
    Brood pheromones ...................................................... 433

# More than Bees ................................................. 435
  Macro and Microfauna ................................................... 435
  Microflora ......................................................................... 435
  Pathogens? ...................................................................... 436
  Upsetting the Balance .................................................... 436
    For More Reading ........................................................ 436

# Bee Math ............................................................... 438

# Races of Bees ..................................................... 439
  Italian ................................................................................ 439
  Starline ............................................................................. 439
  Cordovan .......................................................................... 439
  Caucasian ........................................................................ 440
  Carniolan ......................................................................... 440
  Midnite ............................................................................. 440
  Russian ............................................................................ 441
  Buckfast ........................................................................... 441
  German or English native bees ..................................... 441
  LUS .................................................................................... 442
  Africanized Honey Bees (AHB) ..................................... 442

# Moving Bees ........................................................ 443
  Moving hives two feet .................................................... 443
  Moving hives two miles .................................................. 443
  More than 2 feet and less than 2 miles ....................... 445
  Moving hives 100 yards or less by yourself. ................ 445
    Concepts ...................................................................... 445
    Materials: ..................................................................... 447
    Method ......................................................................... 448

# Treatments for Varroa not working ............ 451

# A Few Good Queens ....................................... 453
  Simple Queen Rearing for a Hobbyist ........................ 453

Labor and Resources.................................................453
Quality of Emergency Queens .................................453
The experts on emergency queens: ..........................454
    Jay Smith:..........................................................454
    C.C. Miller: .......................................................455
Equipment..............................................................456
Method: .................................................................456
    Make sure they are well fed ...............................456
    Make them Queenless...........................................456
    Make up Mating Nucs ..........................................456
    Transfer Queen Cells............................................457
    Check for Eggs....................................................457

# Index to *The Practical Beekeeper* .............. 460

# Volume II Intermediate

# A System of Beekeeping

> "...avoid the mistake of attempting to follow several leaders or systems. Much confusion and annoyance will be saved if he adopts the teachings, methods, and appliances of some one successful beekeeper. He may make the mistake of not choosing the best system, but better this than a mixture of several systems."—W.Z. Hutchinson, Advanced Bee Culture
>
> "In general, the simpler the system, the more efficient and the larger the amount of work which can be accomplished in a given time."—Frank Pellet, Practical Queen Rearing

In this volume, I am going to try to communicate my system of beekeeping. That is not to say it is the only system, but sometimes, as Hutchinson says, mixing up systems may or may not work depending on how well you understand how the parts of that system are related. First let us talk more about systems in general.

## Context

One of the problems in giving beekeeping advice is that we beekeepers tend to give advice based on our system of beekeeping. In other words the advice, by our experience, works in our system of beekeeping. The problem is that this assumes that it will work just as well out of that context and in the context of someone else's system. Sometimes it does. But often it does not.

**Examples**

For example, if my system is to use both upper and lower entrances and a queen excluder and I tell you to wait until you have some bees working the supers to put the excluder on, and your system is to have only a bottom entrance, and you do this, you'll trap a lot of drones in the supers and plug up the excluder with dead drones trying to get out.

Another obvious example would be if I run all the same sized frames and you run deeps for brood and shallows for supers. I tell you the way to get bees working the supers, is to bait them up with a frame of brood, except your brood frames won't fit. Or I tell you to top off their stores by putting some frames of honey in the brood boxes, except your frames of honey are all in shallows and your brood boxes are all deeps.

**Locale**

Local also plays an important role in your system. See the chapter *Locality*. But it seems obvious when talking about cold climates and hot climates. But it does go beyond that as well.

**Summary**

These are simple and obvious but there are many less obvious ones. The fact is that picking and choosing beekeeping techniques from several systems can lead to problems. There is nothing wrong with developing your own system of beekeeping eventually but you need to make sure you learn and understand a system first and know why you are doing what you are doing and then tweak it to meet your needs and your philosophy a little at a time.

## Why a system?

Why do we need a system? Why not just pick and choose what you like? Well, you can, it's just that you have to think about all the ramifications. For instance if you decide you want to do a pollen trap, you have to figure out how the drones will get out. The best ones with the cleanest pollen go on top and that's an adjustment if they are used to a bottom entrance. If you decide to put an excluder on, you have to figure out how the drones on both sides of that excluder will get out. Everything you do has its ramifications and those can affect other things. So that's why we need to work out a system, and not just look at the individual pieces.

## Integration and related issues

## Why this system?

I have designed a system that works for me in my location with my problems. Hopefully you can use it for your situation and your problems. There is nothing wrong with making adjustments to it to fit your style, if you adjust for the ramifications. But here's why I picked the things I did.

## Sustainable

I wanted a system that did not require a lot of input from the outside — Bees in an environment that they could survive without my help.

## Workable

I needed a system that would keep them alive (obviously) and that they could make honey and I could handle the labor involved.

## Efficient

Back to the labor involved, I needed a system that minimized it especially things that were painful or dangerous like lifting really heavy boxes and time consuming things like wiring frames.

# Decisions, Decisions...

*Kinds of beekeeping*

Many decisions depend on what kind of beekeeping you do.

## Commercial

Commercial is generally the term used for someone who does beekeeping as their full-time job. There are different methods of doing this. Usually it involves at least 500 to 1,000 hives.

## Migratory

A migratory beekeeper moves their hives around. Usually they are collecting pollination fees, but sometimes it is just an effort to move south for the winter, so they can build up early and follow the nectar flows north to cash in on as much honey as possible. Pollination is usually something they are paid for.

## Fixed

I'm simply referring to hives that stay in one place for the most part. Usually the beekeeper finds places to put the hives, often not on their own property, where the hives can remain year around. Usually the beekeeper gives some honey to the landowner every fall when the harvest comes in. How much would depend on several things, such as how many hives, how good the forage is for the bees and how much the landowner likes honey. Some just want the bees there, some are hoping for the honey.

## Sideliner

A sideliner is someone with a full-time job already, but they do make some income from the bees. Usually they have from 50 to 200 hives. It's very difficult to keep any number higher and keep a full time job unless you hire some help. It's difficult to make enough money to live on even with 1,000 hives sometimes, so the transition from Sideliner to full-time can be difficult without hired help.

## Hobbyist

A hobbyist is generally defined as anyone who is not making money on the bees. Most hobbyists seem to have about four hives. Two is pretty much the minimum. More than ten or so is a lot of work so most hobbyists tend to stay below that.

### *Personal Beekeeping Philosophy*

A lot of decisions on equipment or methods, depend on your personal philosophy of life and your personal philosophy of beekeeping. Some people have more faith in Nature or the Creator or Evolution to work things out. Some are more interested in keeping their bee healthy with chemicals and treatments. You'll have to decide where you stand on these kinds of things.

## Organic

If you're the type to take an herbal remedy before you run to the doctor, you probably fall into this category. True organic would be no treatments whatsoever. Some will say this can't be done, but there are many people including me doing it. Many are online and help each other through it. After that there are "soft" treatments like essential oils and FGMO, and then slightly "harder" treatments like formic acid and oxalic acid for Varroa.

## Chemical

If you're the type who runs to the doctor for antibiotics the second you get a sniffle this is probably more your style. Some in this group treat for prevention. IMO the wiser ones treat only when necessary. Most of the recent research shows that treating when for prevention has caused resistance to the chemicals on the part of the pests and has done little to help the hive and often hurt them. Chemical buildup in the wax from coumaphos (CheckMite) and fluvalinate (Apistan) used for Varroa mites, is suspected to be the cause of high supersedure rates, and known to be the cause of infertility in drones and queens.

## Science vs. Art

> *"Those who are accustomed to judge by feeling do not understand the process of reasoning, for they would understand at first sight and are not used to seek for principles. And others, on the contrary, who are accustomed to reason from principles, do not at all understand matters of feeling, seeking principles and being unable to see at a glance."—Blaise Pascal*

If you see beekeeping as an art or you see it as a science it will change your perspective a lot. I think it's a bit of both, but since bees are quite capable of surviving on their own and since we really can't coerce them into doing anything, I see it as more of an art where you work with the bees natural tendencies to help them and yourself.

## Scale

This is another thing that changes your philosophy on many things. When you have time to spend with the hives and the hives are in your backyard, then methods that require you to do something every week are not a big problem. For instance, when I requeen in my own yard, I don't mind if it takes three trips to the hive to get it done if that improves acceptance. But if it's at an outyard 60 miles away, I want to do something one time and be done. The same is true of the number of hives. If you have only two hives to deal with on a certain issue, you may not mind how complicated it is. When you have hundreds of hives to deal with, you have to have a streamlined system.

## Reasons for beekeeping

A lot of your decisions will be guided by this. If you have bees as pets you have a different agenda than if you have them solely to make a living.

# Locality

### All Beekeeping is Local

> "In my earlier beekeeping years I was often sorely puzzled at the diametrically opposite views often expressed by the different correspondents for the bee journals. In extension of that state of mind I may say that at that time I did not dream of the wonderful differences of locality in its relation to the management of bees. I saw, measured weighted, compared, and considered all things apicutlural by the standard of my own home—Genesee County, Michigan. It was not until I had seen the fields of New York white with buckwheat, admired the luxuriance of sweet-clover growth in the suburbs of Chicago, followed for miles the great irrigating ditches of Colorado, where they give lift to the royal purple of the alfalfa bloom, and climbed mountains in California, pulling myself up by grasping the sagebrush, that I fully realized the great amount of apicultural meaning stored up in that one little word—locality." —W.Z. Hutchinson, Advanced Bee Culture

It seems rather obvious that beekeeping in Florida won't be the same as beekeeping in Vermont, but what people don't seem to realize is that even in similar winter climates beekeeping is still local. The flows you have in Vermont are not the same as you have in Nebraska. The issues of things like condensation may be very dependent on local climate. For instance, when I was beekeeping in the panhandle of Nebraska, condensation was never a problem. But beekeeping in southeastern Nebraska it is. It's actually colder in the panhandle, and yet, because of differences in humidity, it is not a problem there. All of this seems rather obvious, and yet people continue to ask advice and give advice and contradict advice based on their local experiences without any consideration that warnings given by a beekeeper that they think are unwarranted may be in some locales and not in others. Of course this also applies to things like how many boxes and how much weight do they need to get through the winter and when to manage for swarming and when to start queens and when to do splits and so on.

# Lazy Beekeeping

> *"Everything works if you let it"*—
> *Rick Nielsen of Cheap Trick*

> *"The master accomplishes more and more by doing less and less until finally he accomplishes everything by doing nothing."* —*Laozi, Tao Te Ching*

My grandpa used to say that every great invention came from a lazy man. One of my favorite authors said something similar:

> *"Progress doesn't come from early risers - progress is made by lazy men looking for easier ways to do things."* —*Robert Heinlein*

> *"It's not the daily increase but daily decrease. Hack away at the unessential."*—*Bruce Lee*

In the past few years I've changed most of how I keep bees. Most of it was to make it less work. As of 2007 I've been keeping about two hundred hives with about the same work I used to put into four hives. Here are some of the things I've changed.

## Top Entrances

I've gone to only top entrances. No bottom entrance. I know there are all kinds of people who either hate top entrances or think they cure cancer, or double your honey crop. I don't think either. But I like them and here's why:

1. I never have to worry about the bees not having access to the hive because the grass grew too tall. I also don't have to cut the grass in front of the hives. Less work for me.
2. I never have to worry about the bees not having access because of the snow being too deep (unless it gets over the tops of the hives). So I don't

have to shovel snow after a snowstorm to open the entrances up.

3. I never have to worry about putting mouse guards on or mice getting into the hive.

4. I never have to worry about skunks or opossums eating the bees.

5. Combined with a SBB I have very good ventilation in the summer.

6. I can save money buying (or making) simple migratory style covers. Most of mine are just a piece of plywood with shingle shims for spacers. But some are wider notches in inner covers that I already had.

7. In the winter I don't have to worry about dead bees clogging the bottom entrance.

8. I can put the hive eight inches lower (because I don't have to worry about mice and skunks) and that makes it easier to put that top super on and get it off when it's full.

9. Lower hives blow over less in the wind.

10. This works nicely for long top bar hives when I put supers on because the bees have to go in the super to get in.

11. With some Styrofoam on the top, there's not much condensation with a top entrance in the winter.

Just remember, if you have no bottom entrance and you use an excluder (which I don't) you will need some kind of drone escape on the bottom for them to get out. A $3/8$" hole will do.

More detail in the Chapter *Top Entrances.*

---

***Uniform frame size.***

*"Whatever style (hive) may be adopted, let it by all means be one with movable frames, and have but one sized frame in the apiary."—A.B. Mason, Mysteries of Bee-keeping Explained*

The frame is the basic element of a modern beehive. Even if you have various sized boxes (as far as the number of frames they hold) if the frames are all the same depth you can put them in any of your boxes.

Having a uniform frame size has simplified my life. If all your frames are the same size you have a lot of advantages.

You can put anything currently in the hive anywhere else it's needed.

**For instance:**

1. You can put brood up a box to "bait" the bees up. This is useful even without an excluder (I don't use excluders) but it's especially useful if you really want to use an excluder. A couple of frames of brood above the excluder, leaving the queen and the rest of the brood below, really motivates the bees to cross the excluder and start working the next box above it.

2. You can put honeycombs in for food wherever you need it. I like this for making sure nucs don't starve without the robbing that feeding often starts, or bulking up the stores of a light hive in the fall.

3. You can unclog a brood nest by moving pollen or honey up a box or even a few frames of brood up a box to make room in the brood nest to prevent swarming. If you don't have all the same size, where will you put these frames?

4. You can run an unlimited brood nest with no excluder and if there is brood anywhere you can move it anywhere else. You're not stuck with a bunch of brood in a medium that you can't move down to your deep brood chamber. The advantage of the unlimited brood nest is the queen isn't limited to one or two brood boxes, but can be laying in three or four—probably not four deeps, but probably in four mediums.

I cut all my deeps down to mediums.

Typically I hear the question, "do mediums winter as well?" and I say they winter better in my experience as they have better communication between the frames because of the gap between the boxes. Steve of Brushy Mountain used to say there was some research to this effect, but I'm unsure where to find it.

---

### *Lighter boxes*

> *"Friends don't let friends lift deeps"*
> *—Jim Fischer of Fischer's BeeQuick*

The hardest thing for me about beekeeping is lifting. Boxes full of honey are heavy. Deep boxes full of honey are *very* heavy.

There may be some disagreement as to the exact weights of a full box of honey, and there are other factors involved but in my experience this is a pretty good synopsis of sizes of boxes and typical uses for them:

| 10 frame boxes | | | |
|---|---|---|---|
| Name(s) | Depth | lbs full | Uses |
| Jumbo, Dadant Deep | $11^5/_8"$ | 100-110 | Brood |
| Deep Langstroth Deep | $9^5/_8"$ | 80-90 | Brood & ext |
| Medium, Illinois, $^3/_4$, Western | $6^5/_8"$ | 60-70 | Brood, Ext, Cmb |
| Shallow | $5^3/_4"$, $5^{11}/_{16}"$ | 50-60 | Ext, Cmb |
| Extra Shallow, $^1/_2$ | $4^3/_4"$, $4^{11}/_{16}"$ | 40-50 | Cmb |
| 8 frame boxes | | | |
| Dadant Deep | $11^5/_8"$ | 80-88 | Brood |
| Deep | $9^5/_8"$ | 64-72 | Brood, Ext |
| Medium | $6^5/_8"$ | 48-55 | Brood, Ext, Cmb |
| Shallow | $5^3/_4"$, $5^{11}/_{16}"$ | 40-48 | Ext Cmb |
| Extra Shallow | $4^3/_4"$, $4^{11}/_{16}"$ | 32-40 | Cmb |

If you want a grasp of these and don't have a hive yet, go to the hardware store and stack up two fifty pound boxes of nails or, at the feed store, two fifty pound bags of feed. This is approximately the weight of a full deep. Now take one off and lift one box. This is approximately the weight of a full eight frame medium.

I find I can lift about fifty pounds pretty well, but more is usually a strain that leaves me hurting the next few days. The most versatile size frame is a medium and a box of them that weighs about 50 pounds is an eight frame.

So, first I converted all my deeps into mediums. It was a huge improvement over the occasional deep full of honey I had to lift. I still got tired of lifting 60 pound boxes, so I cut the ten frame mediums down to eight frame mediums and I am really liking them. They are a comfortable weight to lift all day long and not be in pain for the next week. Any lighter and I might be tempted to try to lift two. Any heavier and I'm wishing it was a shade lighter.

I'm wondering how many aging beekeepers have been forced to give up bees because they hurt themselves lifting deeps and it hasn't occurred to them there are other choices?

Richard Taylor in *The Joys of Beekeeping* says:

> *"...no man's back is unbreakable and even beekeepers grow older. When full, a mere shallow super is heavy, weighing forty pounds or more. Deep supers, when filled, are ponderous beyond practical limit."*

I often get asked what the down side of using all eight frame mediums is. There is only one I know of.

8 frame medium vs. 10 frame deep = 1.78 times more initial investment for boxes. ($64 for four eight

frame mediums plus frames vs. $36 for two deeps plus frames)

$512 vs. $288 for eight boxes vs. four boxes
Plus lids and bottoms ($20 either way)
$532 vs. $308 = 1.73 times more or $224
100 hives * $224 = $22,400 which should just about cover your first back surgery.

Typically I hear the question, "do they winter as well?" and I say they winter better in my experience as the cluster fits the box better and they don't leave behind frames of honey on the outside as much as they do in the ten frame hives.

The other big plus is being able to treat a box as a unit when splitting instead of a frame.

More details on how to cut down boxes in Volume three Chapter *Lighter Boxes*.

---

## Horizontal hives

To take not lifting to the next level, how about a hive that's all on one level?

I currently have nine horizontal hives and they have done well. There are some slight adjustments to how to manage them, but the principles are the same. You just can't juggle boxes around. Only frames. But then you can put super on a long hive if you like.

I inherited a few deeps and I already had a Dadant deep, so I currently have three horizontal deeps ($9^5/_8$"), one horizontal Dadant Deep ($11^5/_8$"), four horizontal mediums and one Kenya top bar hive.

I wonder how many old beekeepers, who are being forced to give up their bees, could keep a couple of these without hurting themselves and without much stress?

I wonder how many commercial beekeepers could minimize the labor involved in their operation with these?

I wonder how many hobbyists could just make their life easier with less lifting?

More detail in Volume 3 *Lighter Equipment*.

---

**Top Bar Hive**

Here's another labor saver. How about not even building frames? Or put in foundation—just top bars.

One big long box instead of three separate ones? All the advantages of a horizontal hive. Plus calmer bees because you only face a frame or two of them at a time instead of exposing ten frames of them simultaneously. See Volume 3 *Top Bar Hives* for more detail.

---

***Foundationless frames***

**Making foundationless frames**

You can just break out the wedge on a top bar, turn it sideways and glue and nail it on to make a guide. Or put Popsicle sticks or paint sticks in the groove. Or just cut out the old comb in a drawn wax comb and leave a row at the top or all the way around.

You can cut a triangle off of the corner of a $3/4$" board and have a triangle that on its broad side is $1^1/_{16}$". Or buy some chamfer molding and cut it to length. This can be nailed and glued to the bottom of a top bar to make a peak that the bees will attach to. Once you've made these frames you won't need to put starter strips or foundation in them. Or you can just cut a 45° angle on each side of a top bar before you put the frame together.

Also you can put empty frames with no guides between drawn combs and you can put frames with a top row of cells left on the top bar in anywhere you'd put a frame of foundation.

How much time do you spend putting in foundation, wiring it, tearing it out because it sagged and crumpled or fell out of the frame?

I don't do much of that lately. I mostly use foundationless instead.

And that's not even taking into account the cost of foundation, let alone small cell foundation.

It saves me a lot of work.

Yes, I extract them. I can also use them for cut comb.

No, I don't wire them but you can if you like.

For more detail see the chapter *Foundationless*

---

### No chemicals/no artificial feed

Going to no chemicals saves a lot of work and trouble. All the frames are "clean" so you don't have to worry about residue. If you only feed honey, it's all honey and you don't have to worry what might be syrup instead. You can harvest honey from where ever you find it. And of course you don't have to put in and pull out strips, mix up Fumidil syrup and dust with Terramycin, treat with menthol, make grease patties, fog with FGMO, make up cords, and evaporate oxalic acid. Just think of all the spare time you'll have, and how clean your honey will be.

I've found natural cell size a prerequisite at least for dropping the Varroa mite treatments.

---

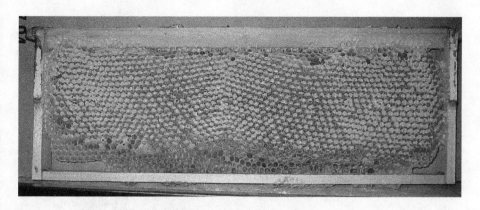

**Leave honey for winter food**

Instead of feeding, just leave them enough. You don't have to harvest it. You don't have to extract it. You don't have to make syrup. You don't have to feed them for winter.

Plus there may be other advantages:

> *"It is well known that improper diet makes one susceptible to disease. Now is it not reasonable to believe that extensive feeding of sugar to bees makes them more susceptible to American Foul Brood and other bee disease? It is known that American Foul Brood is more prevalent in the north than in the south. Why? Is it not because more sugar is fed to bees in the north while here in the south the bees can gather nectar most of the year which makes feeding sugar syrup unnecessary?"—Better Queens, Jay Smith*

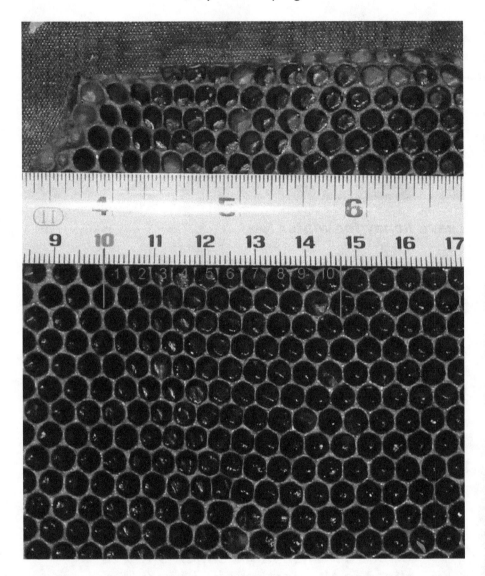

### *Natural cell size*

Of course you get this with foundationless frames or top bar hives, but the "side effect" (or the effect if it's what you were looking for) is not only the labor you save wiring wax or buying and inserting foundation, but once the Varroa mites are under control and your mite counts have stayed stable for a couple of years, you might even be able to forget about Varroa. I have.

It is very nice to be back to just worrying about the bees instead of the mites. See *Natural Cell Size* chapter for more information.

**Carts**

Carts have really helped me with my back. My main yard is across the pasture from my house. Moving boxes, both full and empty, back and forth is a lot of work. It's hardly worth loading the boxes in my van to drive around the long way to get to the hives or vice versa. But it's a long carry. I bought three carts and have used all of them to advantage. I mostly use the Mann Lake and Walter T. Kelley ones right now.

I modified both the Mann Lake and Brushy ones a bit because the boxes would rattle off the cart on the way over to the hives and the Mann Lake one was a little too far off the ground, so I moved the axle up to lower the arms. The Brushy Mt one needed a rack (so they wouldn't rattle off) and a bolt for a stop so I can wheel it around empty. More detail in *Carts*.

**Leave the burr comb between boxes**

> *"Some beekeepers dismantle every hive and scrape every frame, which is pointless as the bees soon glue everything back the way it was." — The How-To-Do-It book of Beekeeping, Richard Taylor*

Here's one I think helps the bees, gives you a chance to monitor for mites on drone pupae and saves work. Leave the burr comb that goes from the bottom of one frame to the top of the one below it. Yes it will break when you separate the boxes, but it makes a nice ladder for the queen to get from one box to the next. Also, they often build some drone comb between the boxes and if you tear them open you'll see the drone pupae and maybe you'll notice mites (you should be looking).

## Stop cutting out swarm cells

I read the books and I tried to do this when I was young, inexperienced and foolish. The bees soon taught me what a waste of time and effort it was. If the bees have made up their mind to swarm, do a split or put each frame with some swarm cells in a nuc with a frame of honey and get some nice queens. Once they've gone this far, I've never seen them change their mind. Of course the solution was to keep it from getting this far. Keeping the brood nest open while keeping enough expansion room in the supers is the best swarm control I've found. If the brood nest is getting filled with honey, put a couple of empty frames in. Yes, empty. No foundation, nothing. Try it. The bees will build some drone comb, probably the first frame, but after that they'll draw some very nice worker brood and the queen will have it laid up before the whole comb is even drawn or even full depth. You'll be shocked how quickly they can do this and how it distracts them from swarming.

### Stop fighting your bees

> *"There are a few rules of thumb that are useful guides. One is that when you are confronted with some problem in the apiary and you do not know what to do, then do nothing. Matters are seldom made worse by doing nothing and are often made much worse by inept intervention."*
> *—The How-To-Do-It Book of Beekeeping, Richard Taylor*

I don't know how often I see questions on bee forums asking how can I make the bees do this or that. Well, you can't make them do anything. In the end they do what bees do no matter what you try to make them do. You can help them out, by making sure they have the resources they need to do what you think they need to do and by manipulating the hive so they don't swarm. You can fool them into making queens and such. But you'll have a lot more fun and work a lot less if you stop trying to make them do anything.

### Stop wrapping your hive.

> *"Although we now and again have to put up with exceptionally severe winters even here in the southwest, we do not provide our colonies with any additional protection. We know that cold, even severe cold, does not harm colonies that*

> are in good health. Indeed, cold seems to have a decided beneficial effect on bees." —*Beekeeping at Buckfast Abbey*, Brother Adam
>
> "Nothing has been said of providing warmth to the colonies, by wrapping or packing hives or otherwise, and rightly so. If not properly done, wrapping or packing can be disastrous, creating what amounts to a damp tomb for the colony" —*The How-To-Do-It Book of Beekeeping*, Richard Taylor

I suppose this also includes all the worrying about winter and trying to give them heaters and such. The bees have lived for millions of years with no heaters and no help. If you make sure they are strong and have enough food and adequate ventilation so they don't end up in an icicle from condensation, then you should relax. Work on your equipment and see them in the spring, or at the earliest, late winter.

## Stop scraping all the propolis off of everything

> "Propolis rarely creates problems for a beekeeper. Certainly any effort to keep a hive free of it by systematic and frequent scraping, is time wasted." —*The How-To-Do-It Book of Beekeeping*, Richard Taylor

Doesn't it feel like a losing battle anyway? The bees will just replace it, so unless it's directly in your way, why bother?

---

***Stop painting your equipment.***

*"The hives need no painting, although there is no harm in doing it if their owner wants to please his own eye. The bees find their way to their own hives more easily if the hives do not all look alike. I rarely paint mine, and as a result no two are quite alike. Most have the appearance of many years of use and many seasons of exposure to the ele-*

> ments." —Richard Taylor, The Joys of Beekeeping

> "I suppose they would last longer if painted, but hardly enough longer to pay for the paint." —C.C. Miller, Fifty Years Among the Bees

You've probably noticed by now, if you looked at pictures of my hives, that a lot of them are not painted. Maybe the neighbors or the wife will complain but the bees won't care. They might not last as long. I don't know because I only stopped painting them about four years ago. But think of all the time you'll save!

Lately I bought a lot of equipment and wanted to keep it as nice as I could for as long as I could so I started dipping them in beeswax and gum rosin.

### Stop switching hive bodies.

> "Some beekeepers, trusting the ways of bees less than I do, at this point routinely 'switch hive bodies,' that is, switch the positions of the two stories of each hive, thinking that this will induce the queen to increase her egg laying and distribute it more widely through the hive. I doubt, however, that any such result is accomplished, and in any case I have long since found that such planning is best left to the bees." –Richard Taylor, The Joys of Beekeeping

In my opinion switching hive bodies is counterproductive. It's a lot of work for the beekeeper and it's a lot of work for the bees. After you swap them the bees have to rearrange the brood nest. It's true it will interrupt swarming, but so will other things. See the chapter *Swarm Control* for what I do.

---

**Don't look for the queen.**

Don't look for the queen unless you have to. It's one of the most time consuming operations. Instead look for eggs or open brood. Nothing wrong with keeping your eye out for her, but trying to find her is time consuming. This even works for things like setting up mating nucs. If you break up a hive for mating nucs and don't look for the queen on the frames and give to the nucs you may lose a queen, but you'll save a lot of time. She'll just get superseded. The only real advantage to finding the queen often is the practice but this could be more easily done with an observation hive.

If you have issues you are concerned about regarding queens, give them a frame of eggs and open brood from another hive and move on. If they are queenless they will raise one. If they are not, you haven't interfered. See the Volume I BLUF the section Panacea for more information.

**Don't wait.**

There are many operations where people, including me, will tell you to remove the queen and wait until the next day. This would be things like introducing queen cells to nucs or introducing a new queen to a hive. Waiting will improve the odds of acceptance, But reality is it will only improve it a little. So if you want to save time, don't wait until the next day unless you have to, do it now while you have the hive open.

### Feed dry sugar.

No, they won't take it as well, but if you have to feed it will keep them from starving and you won't have to make syrup and you won't have to buy feeders and you won't have any drowned bees. See *Feeding Bees* for more details.

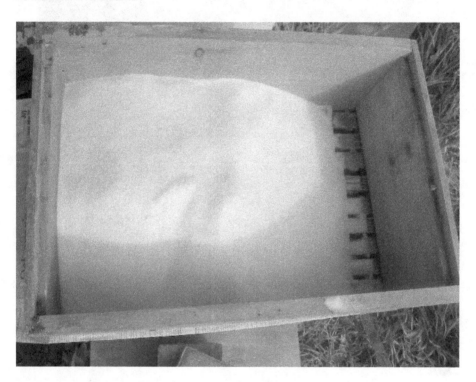

---

### Split by the box.

If you've got a booming hive you want to split in the spring, don't look for the queen, don't look for brood, just split it by boxes. The bottom two boxes that are seriously occupied by bees probably have brood in them. Of course success is mostly dependent on being able to guess pretty accurately that you have brood and stores in both boxes. If you're wrong, you'll end up with one box empty after only a day or so. But if you are right, you've saved a lot of work. With eight frame mediums (which are half the volume of a ten frame deep) the odds of this working on a hive that is at least four boxes (the equivalent of two ten frame deeps) is twice as good. You just deal the boxes like cards. Put a bottom board on each side and do "one for you and one for you" until you're done. Come back in a month and see how they are doing.

---

**Stop Requeening.**

If you let the bees requeen themselves you'll breed bees that *can* and *do* requeen themselves. Bees in nature have this selective pressure on them. Bees that are constantly requeened by the beekeeper do not have this selective pressure on them. I would only requeen if the hive seems to be failing and I would do so from a hive that is successful at requeening themselves.

Along with this, of course, stop buying queens. Make splits and let the bees raise their own. That way you get bees that are well adapted to your climate and your pests and your diseases; *and* you get diseases and pests that are well adapted to coexist with the bees instead of killing them.

# Feeding Bees

You would think something this simple would not be controversial, but it is—on several fronts.

**First, when do you feed?**

> "Q. When is the best time to feed the bees?
>
> "A. The best thing is never to feed them, but let them gather their own stores. But if the season is a failure, as it is some years in most places, then you must feed. The best time for that is just as soon as you know they will need feeding for winter; say in August or September. October does very well, however, and even if you haven't fed until December, better feed then than to let the bees starve."
>
> —C.C. Miller, A Thousand Answers to Beekeeping Questions, 1917

In my opinion there are many reasons to avoid feeding if you can. It sets off robbing. It attracts pests (ants, wasps, yellow jackets etc.) It clogs the brood nest and sets off swarming. It drowns a lot of bees, not to mention it's a lot of work. Then if you use syrup there is the effect of the pH on the microbial culture of the hive and difference in nutritional value compared to what they would have gathered on their own.

Some people feed a package constantly for the first year. In my experience this usually results in them swarming when they are not strong enough and often failing. Some feed spring, fall and dearth regardless of stores. Some don't believe in feeding at all. Some steal all the honey in the fall and try to feed them back up enough to winter.

Personally I don't feed if there is a nectar flow and they have some capped stores. Gathering nectar is what bees do. They should be encouraged to do it. I will feed in the spring if they are light, as they will not rear brood without sufficient stores to do it with. I will feed in the fall if they are light, but I always try to make sure I don't take too much honey and leave them light. Some years, though, the fall flow fails and they are on the verge of starvation if I don't feed. When queen rearing, during a dearth, I sometimes have to feed to get them to make cells and to get the queens to fly out and mate. So while I do try to avoid feeding, I end up doing it very often. In my opinion, there is nothing wrong with feeding if you have a good reason for doing it, but my plan is to try to avoid it and leave the bees enough to live on. Also, while I think honey is the best food for them, it's too much work to harvest it and then feed it back, so when I feed it's either dry sugar or sugar syrup, unless I have some honey I don't think is marketable.

Pollen, if fed, is usually fed before the first available pollen in the spring. Here (Greenwood, Nebraska) that would be about mid February. I have not had luck getting bees to take it any other time except a fall dearth.

## Stimulative feeding.

A lot of literature out there will act like stimulative feeding is an absolute necessity to get honey production. Many of the greats of beekeeping have decided this is not productive:

> *"The reader will by now have drawn the conclusion that stimulative feeding, apart from getting the foundations drawn out in the brood chamber, plays no part in our scheme of bee-keeping. This is in fact so." —Beekeeping at Buckfast Abbey, Brother Adam*

> *"Very many, at the present time, seem to think that brood rearing can be made to forge ahead much faster by feeding the bees a teacupful of thin sweet every day than by any other method; but from many experiments along this line during the past thirty years I can only think this a mistaken idea, based on theory rather than on a practical solution of the matter by taking a certain number of colonies in the same apiary, feeding half of them while the other half are left "rich" in stores, as above, but without feeding and then comparing "notes" regarding each half, thus determining which is the better to go into the honey harvest...results show that the "millions of honey at our house" plan*

*followed by what is to come hereafter, will outstrip any of the heretofore known stimulating plans by far in the race for bees in time for the harvest."* —A Year's work in an Out Apiary, G.M. Doolittle.

*"Probably the single most important step in management for achieving colony strength, and one most neglected by beekeepers, is to make sure the hives are heavy with stores in the fall, so that they emerge from overwintering already strong early in the spring"* —The How-To-Do-It Book of Beekeeping, Richard Taylor

*"The feeding of bees for stimulating brood-rearing in early spring is now looked upon by many as of doubtful value. Especially is this true in the Northern States, where weeks of warm weather are often followed by 'Freeze up.' The average beekeeper in the average locality will find it more satisfactory to feed liberally in the fall— enough, at least so that there shall be sufficient stores until harvest. If the hives are well protected, and the bees well supplied with an abundance of sealed stores, natural brood rearing will proceed with sufficient rapidity, early in the spring without any artificial stimulus. The only time that spring feeding is advisable is where there is a*

> *dearth of nectar after the early spring flow and before the coming of the main harvest."* —W.Z. Hutchinson, Advanced Bee Culture

**My experiences with stimulative feeding.**

I've tried about every combination over the years and my conclusion is that weather has everything to do with the success or failure of any stimulative feeding attempt. So some years it seems to help some, some years it misleads them into rearing too much brood too early when a hard freeze could be disastrous or having too much moisture in the hive in that precarious time of late winter when a hard freeze could still happen. Plus the really impressive results you get are usually from feeding a hive that is light in stores. Leaving more stores still seems to be a more reliable method of getting a lot of early brood in my climate.

Here in the North it not only makes it difficult to even do, but makes the results vary from disastrous to remarkable. The problem is that beekeeping has enough variables and I'm not interested in introducing more.

I will skip the what to feed issues and distill them down to my experience as relates to stimulating brood production and ignore the issues of honey vs. sugar for the moment.

I have fed really thin (1:2) thin (1:1) moderate (3:2) and thick (2:1) syrup at every time of the year except a honey flow, but again to simplify the issue to stimulating brood rearing, let's stick with the spring.

I see no difference in brood stimulation between any of the ratios. The bees will suck it down if it's warm enough (and here it seldom is in early spring or late fall) and it will induce them sometimes to start brood rearing when the bee's common sense is that it is too early. So for simplifying even further, let's just talk about feeding or not feeding syrup.

Difficulty getting bees to take syrup early in Northern climates:

If you try to feed any kind of syrup to bees in my climate in the late winter or early spring, the results *usually* are that they will not take it. The reason is that the syrup is hardly ever above 50º F (10º C). At night it is somewhere between freezing and sub zero. In the daytimes it's usually not above freezing on those rare occasions when it's actually 50º F in the daytime, the syrup is still below 32º F (0º C) from the night before. So first of all, trying to feed syrup in the late winter and early spring usually doesn't work at all—meaning they won't even take it.

## Down sides to success:

Then, if you get lucky and get some warm spell somewhere in there that stays warm enough long enough for the syrup to get warm enough that the bees will take it, you manage to get them rearing a huge amount of brood, let's say near the end of February or early March, and then you get a sudden sub zero freeze that lasts for a week and all of the hives that were so induced to raise brood, die trying to maintain that brood. They die because they won't leave it and they die because they can't keep it warm, but they try anyway. We could get a hard freeze (10º or below F) anywhere up to the end of April, and last year we did get one in mid April as did most of the country.

Our record low, here in the warmest part of Nebraska in February is -25º F. In March it's -19º F. In April it's 3º F (16º C). In May it's 25º F (-31º C). Having freezing weather in May is common enough here.

I've seen snowstorms on May 1st. So I seriously doubt, not only the efficacy of feeding syrup, but if you can get it to work, the wisdom of stimulating brood rearing ahead of what is normal for the bees anyway.

**Variable outcomes:**

This might be an entirely different outcome in one year than another year. Certainly if your gamble pays off and you get the bees to brood up in March and you manage to keep them from swarming in April or May (doubtful), don't get any hard freezes that kill some of the hives off, or they are built up so far by the time those freezes hit that they can manage, and you manage to keep that max population for the flow in mid June, maybe you'll get a bumper crop. On the other hand, you get them to brood up heavy in March, get a subzero freeze that lasts a week and most of them die, it's a very different outcome.

In a different climate, this might be an entirely different undertaking. If you live where subzero is unheard of, and clusters don't get stuck on brood from cold and can't get to stores, then the results of stimulative feeding may be much more predictable and possibly much more positive. Then again they may brood up too early and swarm before the flow.

**Dry Sugar:**

This is not good spring feed, except as left over from winter, but in my experience it made a lot of difference overwinter and in the following spring. Most of the hives ate the sugar. Some ate most of the sugar. They did brood up while eating sugar and they could eat it even when it was cold. They don't go as crazy over it nor as crazy on brood rearing, but I see that as a good thing. A moderate build up from stores they can get at even in the cold is a much better survival bet than a huge build up at a time they could get caught in long hard freeze on syrup that they won't be able to get to if it's cold.

**Type of feeder:**

I will admit, that the type of feeder also plays into all of this. A top feeder in the early spring here is worthless. The syrup is hardly ever warm enough for the bees to take it. Baggie feeders, on the other hand, on top of the cluster, they seem to be able to get at, as well as dry sugar. A frame feeder (as much as I don't like them) against the cluster is taken much better than the top feeder. (but not as well as the baggie feeders). In my climate any feeder that is very far from the cluster will not get used until the weather is consistently in the 50's F (10s C) and by then the fruit trees and dandelions will be blooming so it will be irrelevant.

You might get some syrup down them in late March or early April with a baggie feeder or a jar or pail directly over the cluster or if you reheat the syrup regularly, when everything else fails.

*Second,* **what** *do you feed?*

I prefer to *leave* them honey. Some think you should only feed honey. From a perfectionist view, I like that. From a practical view, it's difficult for me. First, honey sets off robbing a lot worse than syrup. Second, honey spoils a lot more easily if I water it down, and I hate to see honey go to waste. Third, honey is very expensive (if you buy it or just don't sell it) and labor intensive to extract it. It seems wrong to me to go to the trouble of extracting it, only to feed it back. I'd rather leave enough honey on the hives and, in a pinch, steal some from a stronger hive for the weaker hives, rather than feeding. But if it comes down to needing to feed, I feed off, old, or crystallized honey if I have it, otherwise I feed sugar syrup.

## Pollen

The other issue of what, of course, is pollen and substitute. The bees are healthier on real pollen, but substitute is cheap. I try to feed all real pollen, but sometimes I can't afford that and I settle for 50:50 pollen:substitute. On just substitute you get very short-lived bees. I don't notice any difference at 50:50, but I still think 100% pollen is best.

### Third, how much do you feed?

It's best to check with local beekeepers on how much stores it takes to get through your winters. Here, with a large cluster of Italians, I'd shoot for a hive weight of 100 to 150 pounds. With Carniolans, it's more like 75 to 100 pounds. With the more frugal feral survivors it might be more like 50 to 75 pounds. It's always better to have too much than too little.

### Fourth, how do you feed?

There are more schemes to feed bees than there are options in any other aspect of beekeeping. I have a love/hate relationship with feeding to start with so it's not surprising I have a love/hate relationship with most methods.

***Issues when considering the type of feeder***:

How much labor is involved in feeding? For instance do I have to suit up? Open the hive? Remove lids? Remove boxes? How much syrup will it hold? How many trips will I have to make to an outyard to get them ready for winter? In other words, a feeder that holds five gallons of syrup, I'll only have to fill once. If it only holds a pint or a quart I'll have to fill it many times.

Will the bees take it if it's cold? If the weather is warm most any feeder works. Only a few will work when the weather is marginal. Meaning it's in the 40's or so at night and the 50's or so in the day and none work when it stays too cold all the time.

What does it cost? Some methods are quite expensive (a good hive top feeder could cost $20 to $40 per hive) and some are quite cheap (converting a solid bottom board to a feeder might cost 25¢ per hive).

Does it cause robbing? Boardman feeders, for example are notorious for this.

Does it cause drowning? Can this be mitigated? Frame feeders are notorious for this and most beekeepers have added a float or ladder or both to minimize it. Bottom board feeders are about the same as the frame feeders.

Is it hard to get into the hive with the feeder on or does it get in the way? For instance a top feeder has to be removed to get into the hive and it sloshes and spills a lot.

Is it hard to clean out the feeder? Feed will spoil. Feeders will get mold in them. If bees can drown in them, they will have to be cleaned out from time to time.

## Basic types of feeders

### Frame feeders

Frame feeder. These vary a lot. The really old ones were wood. The old ones were smooth plastic and drowned a lot of bees. The newer ones are mostly a black plastic trough with some roughness on the sides to act as a ladder. If you put a float in them they work much better with less drowning or a #8 hardware cloth ladder helps. They also take up more than one frame, more like a frame and a half so they don't fit well and they bulge in the middle. Brushy used to have one made one out of Masonite with more limited access, a built in #8 ladder and it only takes up one space and it doesn't bulge. Betterbee has a plastic version with similar features. I haven't had one, but the complaints I've heard are that the ears are too short and it falls off the frame rest. If you make them correctly then they would live up to their other name "division board feeder", but to do that they have to divide the hive into two parts and should have separate access for each side of the hive. Some people make actual "division board feeders" themselves and use them to make a ten frame hive into two four frame nucs with a shared feeder.

## Boardman feeder

These come in all the beginners' kits. They go in the entrance and hold an inverted quart mason jar. I'd keep the jar lid and throw away the feeder. They are notorious for causing robbing. They are easy to check but you have to shake off the bees and open the jar to refill them.

## Jar feeder

Inverted container. These work on the same principle as a water cooler or other upside down containers where the liquid is held in by a vacuum (or for the technically minded among us, held in by the air pressure outside pushing on it). For feeding bees, this can be a quart jar (like the one from the Boardman feeder), a paint can with holes, a plastic pail with a lid, a one liter bottle etc.

It just has to have some way to hold it over the bees and some small holes for the syrup to get out. Advantages vary by how you set it up and how big they are. If they hold a gallon or more you won't have to refill very often. If they are only a quart you will have to refill a lot. If they leak or the temperature changes a lot, they leak and drown or "freeze" the bees. They are usually cheap and usually drown fewer bees than the frame feeder, unless they leak. If the hole it goes over is covered in #8 hardware cloth you won't have any bees on the container when you need to refill.

## Miller feeder

Named after C.C. Miller. There are variations of this. All go on top of the hive and require tight closure so robbers don't get in the top and drown in the syrup. Some of them have open access by the bees to the entire feeder. Some have limited access that is screened in so the bees have just enough room to get to the syrup. They come with the access in various places—sometimes one end, sometimes both, sometimes the center parallel to the frames and sometimes across the frames. The reasoning is either based on being easier to make and fill with only one compartment (ends) or better access for the bees (center) or even better access for the bees (across the frames) so the bees will find it. The taller they are the less they get used when it's cold but the more syrup they hold. Some hold as much as five gallons (great for an outyard during warm weather but not good when it's cold at night). Some hold as little as a couple of quarts. For cool weather the bees will work one that is shallow and has the entrance in the center better than one that is deep and has the entrance on the end. The Rapid feeder is a similar concept but is round and goes over the inner cover hole. The biggest disadvantage is probably having to remove it to get into the hive. Pretty awkward if it's full. The biggest advantages are the volume of syrup they hold and (if it's screened) filling without having to suit up or disrupt the bees.

## Bottom board feeder

### Jay Smith Bottom Board Feeder

Jay Smith Bottom Board Feeder is simply a dam made with a $^3/_4"$ by $^3/_4"$ block of wood put an inch or so back from the where the front of the hive would be (18" or so forward of the very back). The box is slid forward enough to make a gap at the back. The syrup is poured in the back. A small board can be used to block the opening in the back. The bees can still get out the front by simply coming down forward of the dam. The picture is from the perspective of standing behind the hive looking toward the front. This is all empty so you can see where the dam is etc. The edges of the dam have been enhanced and labels put on to try to make more sense. This version doesn't work on a weak hive as the syrup is too close to the entrance. It drowns as many bees as the frame feeders.

*Jay Smith Bottom Board Feeder*

## My version

*Bottom of the feeder. The block part way across makes a reduced entrance for the hive below it.*

Top of the feeder. The dam at the front stops the syrup from running out. The support block holds the #8 hardware cloth up so it doesn't sag. The #8 lets me fill the feeder without bees flying out. The drain plug is so I can let condensation out in the winter or rain water if it gets in. It's been dipped in wax and the cracks filled with a wax tube fastener. You could just melt some beeswax and roll it around in the feeder to seal it.

With a box on it so you can see where you fill it. If you aren't stacking them the feeder portion could be on the front or back. When doing "apartment style" the filler is in the front.

*Apartment style where you can see the entrance for the nuc below on the bottom.*

Apartment style with covers over the filler to keep out most of the rain. These are scraps of $1/2$" plywood, but anything works fine. So far they haven't blown off.

My version of the Jay Smith Bottom Board Feeder I just modified this to make a top entrance and a bottom feeder. These were made from a standard bottom board from Miller Bee Supply. The space on top is $3/4$" and the space on the bottom is $1/2$". This is a nice space for overwintering as I can put some newspaper on and cover with sugar, or I can fit a pollen patty in without squishing bees. I was concerned about water from condensation so I added a drain plug. This could also be used to drain bad syrup. Also this design allows stacking up nucs and feeding all of them without opening or rearranging. So far I have had about the same number of drownings as a standard frame feeder. You do have to pour the syrup in slowly and if the bees are obviously so thick that they are all over the bottom you might want to add a box and lessen the congestion. I am considering making a float out of $1/4$" luan.

## Baggie feeder

These are just gallon Ziploc baggies that are filled with three quarts of syrup, laid on the top bars and slit on top with a razor blade with two or three small slits. The bees suck down the syrup until the bag is empty. A box of some kind is required to make room. An upside down Miller feeder or a one by three shim or just any empty super will work. Advantages are the cost (just the cost of the bags) and the bees will work it in cooler weather as the cluster keeps it warm. Disadvantages are you have to disrupt the bees to put new bags on and the old bags are ruined. Also the risk of excess space in the hive that could get comb in it.

## Open feeder

These are just large containers with floats ("popcorn peanuts", straw etc.) full of syrup. They are usually kept away from the hives a ways (100 yards or more). Advantages are you can feed quickly as you don't have to go to every hive. Disadvantages are that you are feeding the neighbor's bees and they sometimes set off robbing and sometimes in a feeding frenzy a lot of bees drown.

## Candy board

This is a one by three box with a lid that has candy poured into it. It goes on top in the winter and the bees will use it if they get to the top of the hive and need food. They are very popular around here and seem to work well.

## Fondant

This can be put on the top bars. Again it seems to be more useful for emergency feed. The bees will eat it if there is nothing to eat. The end effect is similar to the candy board.

## Dry Sugar

This can be fed a number of ways. Some people just dump it down the back of the hive (definitely not recommended with screened bottom boards as it will fall through to the ground). Some put it on top of the inner cover. Some put a sheet of newspaper on top of the top bars, add a box on top and put the sugar on the newspaper (as in the photos above). Others put it in a frame feeder (the black plastic trough kind). I've even pulled two frames out of an eight frame box that were empty and dumped the sugar in the gap (with a solid bottom board of course). With screened bottom boards or with a small hive that just needs a little help, I'll pull some empty frames out, put some newspaper in the gap and put a little sugar, spray a little water to clump it so it doesn't run out, a little more sugar until I get it full. Sometimes the house bees carry it out for trash if you don't clump it. If you drizzle some water on it you can get the bees interested in it. The finer the sugar the better they take it. If you can get "bakers" sugar or "drivert" sugar it will be better accepted that standard sugar but harder to find and more expensive.

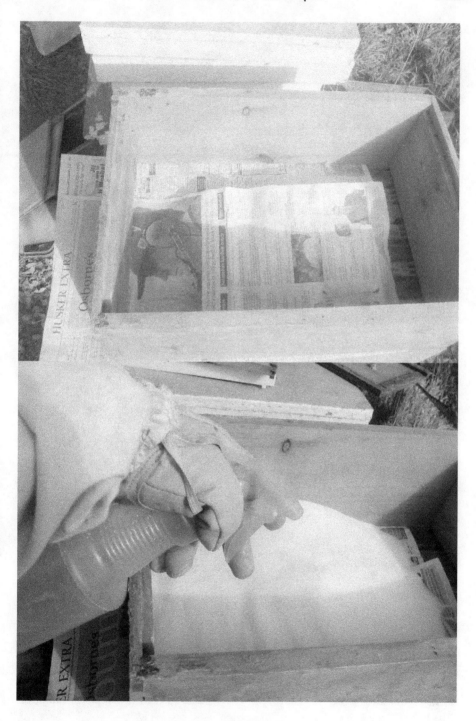

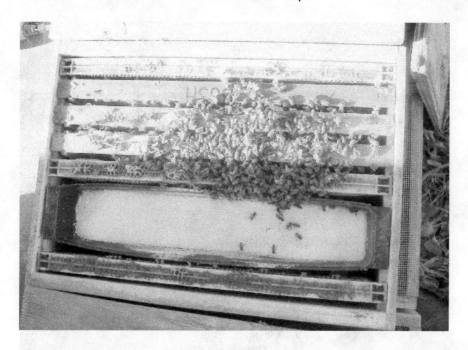

## What kind of sugar?

It matters not at all if it's beet sugar or cane sugar.

It matters a lot if it's granulated white sugar or anything else. Powdered sugar, brown sugar, molasses and any other unrefined sugar is not good for bees. They can't handle the solids.

## Pollen

Pollen is fed either in open feeders for the bees to gather it (dry) or in patties (mixed with syrup or honey into a dough and pressed between sheets of waxed paper). The patties are put on the top bars. A shim is helpful to make room for the patty. I usually do open feeding dry in an empty hive on screen wire on top of a solid bottom so it won't mold.

## Measuring ratios for syrup

The standard mixtures are 1:1 in the spring and 2:1 in the fall (sugar:water). People often use something other than those for their own reasons. Some people use 2:1 in the spring because it's easier to haul around and keeps better. Some people use 1:1 in the fall because they believe it stimulates brood rearing and they want to be sure to have young bees going into winter. The bees will manage either way. I use more like 5:3 (sugar:water) all the time. It keeps better than 1:1 and is easier to dissolve than 2:1.

## Weight or Volume?

The next argument is over weight or volume. If you have a good scale you can find this out for yourself, but take a pint container, tare it (weigh it empty) and fill it with water. The water will weigh very close to a pound. Now take a dry pint container, tare it (weigh it empty) and fill it with white sugar and weight it. It will weigh very close to a pound. So I'll keep this very simple. For the sake of mixing syrup for feeding bees, it just doesn't matter. You can mix and match. "A pints a pound the world around" as far as dry white sugar and water are concerned. At least until you've mixed the syrup. So if you take 10 pints of water, boil it, and add 10 pounds of sugar you'll get the same thing as if you took 10 pounds of water, boil it, and add 10 pints of sugar.

The next confusion seems to be on how much it takes to make how much syrup. The volume of 10 pints of water and 10 pints of sugar will make about 15 pints of syrup, not 20. The sugar and the water fit together.

**How to measure**

Don't confuse the issue of how you measure. Measure before you mix. In other words, you can't fill a container $1/3$ of the way with water, and add sugar until it's $2/3$ full and have 1:1 syrup. You'll get more like 2:1 syrup. Likewise, you can't fill it $1/3$ of the way with sugar and then add water until it's $2/3$ full and have 1:1 syrup. You'll get more like 1:2. You have to measure both separately and then put them together to get an accurate measurement. I find the easiest is to use pints for water and pounds for sugar since the sugar comes in packages marked in pounds and volume is easy to measure for water. So if you know you are going to add 10 pounds of sugar and you want 1:1 then start with 10 pints of boiling water and add the 10 pounds of sugar.

**How to make syrup**

I boil the water and add the sugar and then when it's all dissolved turn off the heat. With 2:1 this can take /some time. Either way, boiling the water makes the syrup keep longer by killing all the microorganisms that might be in the sugar or the water.

## Moldy syrup

I don't let a little mold bother me, but if it smells too funny or it's too moldy I throw it out. If you use essential oils (and I don't) they tend to keep it from molding. Some people add various things to control this. Clorox, distilled vinegar, vitamin C, lemon juice and other things are used by various people to help it keep longer. All of these except the Clorox make the syrup more acidic and closer to the acidity of honey (lower the pH).

# Top Entrances

### Reasons for top entrances

You can keep bees fine without these, but they do eliminate the following problems: mice, skunks, opossums, dead bees blocking the exit in winter, condensation on the lid in winter, snow blocking the exit in winter, grass blocking the exit the rest of the year. It also allows you to buy inexpensive and very nice Sundance II pollen traps.

> *"I had a neighbor who used the common box hive; he had a two inch hole in the top which he left open all winter; the hives setting on top of hemlock stumps without any protection, summer or winter, except something to keep the rain out and snow from beating into the top of the hive. he plastered up tight all around the bottom of the hive for winter. his bees wintered well, and would every season swarm from two to three weeks earlier than mine; scarcely any of them would come out on the snow until the weather was warm enough for them to get back into the hive.*
>
> *"Since then I have observed that whenever I have found a swarm in the woods where the hollow was below the entrance, the comb was always bright and clean, and the*

*bees were always in the best condition; no dead bees in the bottom of the log; and on the contrary when I have found a tree where the entrance was below the hollow, there was always more or less mouldy comb, dead bees &c.*

*"Again if you see a box hive with a crack in it from top to bottom large enough to put your fingers in, the bees are all right in nine cases out of ten. The conclusion I have come to is this, that with upward ventilation without any current of air from the bottom of the hive, your bees will winter well..."*—Elisha Gallup, The American Bee Journal 1868, Volume 3, Number 8 pg 154

332                    Top Entrances

*Regular migratory covers with tapered shims to make top entrances with the opening the long way.*

**How to make top entrances**

My current ones are these. These are $3/4"$ plywood cut to the size of the box (no overhang or cleats) with shims to make the opening the short way.

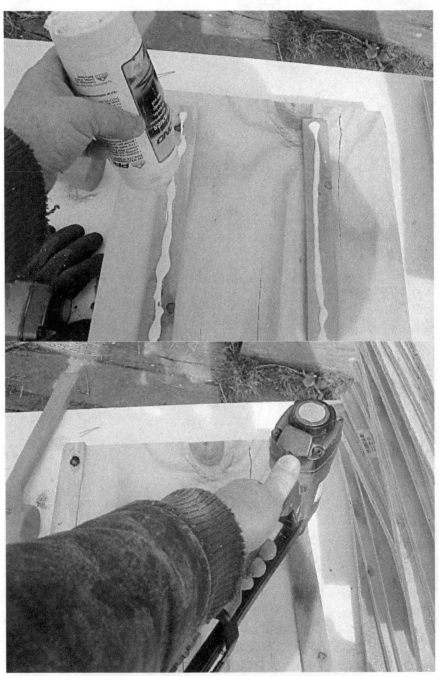

*Making the top entrances.*

I recently started making them out of $1/2"$ plywood.

The idea of using shims was presented to me by Lloyd Spears who says he got it from a man named Ludewig

**Top Entrance Frequently Asked Questions:**

**Q:** Without a bottom entrance, don't they have trouble hauling out the dead bees and keeping the hive clean?

**A:** In my observation, no more than with a bottom entrance. Either way dead bees accumulate over winter. Either way they accumulate some in the fall. Either way they usually keep it pretty clean in the middle of the year. I've watched a house keeping bee in my observation hive (which has a bottom entrance), haul dead bees all over the hive from top to bottom before finally finding the entrance at the bottom. I don't think it matters at all. According to Elisha Gallup (see previous quote) the opposite is true. He says the upper entrance ones are clean of debris while the bottom entrance ones are full of debris.

**Q:** Do the returning foragers get irate when you're working the hive?

**A:** I haven't noticed any difference. Whether a top entrance or a bottom entrance, while you're working the hive you're disrupting things just by standing there. You always, in both cases, have confused bees circling and with both bottom and top entrances you have bees who just go back into the hive while you're working. With the top entrance they just go in the top.

**Q:** When removing supers don't they get confused?

**A:** The most confusion is when you remove them from only one and it's right next to a similar height hive. Then they do get confused about which hive is theirs. But I think they do the same with a bottom entrance for the same reason except you don't notice. They use the height of the hive as one of their landmarks so they continue to fly into the tall white hive nearest where they remember it instead of the short one next to it. In a day things go back to normal.

**Q:** Why do some people recommend not using them in town because of bees being confused when working the hive?

**A:** Similar to above answer. In my experience any hive being opened causes confusion for the returning foragers because the height of the hive is often changed because of removing boxes, and the beekeeper's presence changes the landmarks. I see no increase in the confusion of a top entrance only hive to a bottom entrance only hive. In my opinion, advice that top entrances should not be used in urban areas are misplaced but seem to be often repeated by those who have no experience with top entrances. Wintering will be much improved by a top entrance and it prevents issues such as and overheating as well. These advantages should not be sacrificed merely because of a commonly held belief about hive disruption that is so often repeated.

**Q:** Do you use an entrance reducer?

**A:** On some of them I have them and some I don't. I use a $1/4$" thick piece of wood (a piece of screen molding works well) cut 2" short of the width of the opening with one nail in the center to make a pivot so you can pivot it open or pivot it closed.

# Carts

In my pursuit of easier beekeeping, I bought and modified these carts.

I modified two of the beekeeping carts I have. This is the Brushy Mt one. I added the perforated angle iron rack on the front so I can haul six empty boxes around without them sliding off. I also added the bolt to the stop so I can move it when it's empty. Unfortunately I'll have to drill another hole for the pin if I want to haul 8 frame boxes with it.

Here's the rack on the Mann Lake beekeeping cart. Again, so I can haul six empty boxes across the pasture without them falling off. The pin in the hole at the top is used too, to keep the boxes from tipping forward when you pick them up. I had to lower the axle by adding the angle iron on top here so it would slip into a medium and pick it up without fighting with it tipping forward. I also had to cut off some of the angle iron on the bottom so it wouldn't catch in the grass. I seem to use this one the most because you can just slide into a stack of boxes and pick them up.

This one, by the way, was invented by beekeeper Jerry Hosterman of Arizona. I've seen some of his work that are obviously much older than Mann Lake's.

Here's the classic Walter T. Kelley "Nose Truck" designed for beekeeping. It requires some kind of bottom board, preferably with some cleats on the end, to act as a pallet. It's heavy duty and will haul six FULL supers. I did no modifications on it.

# Swarm Control

*Photo by Judy Lillie*

Swarming is when the old queen and part of the bees leave to start a new colony. Afterswarms are after the old queen has left and there are still too many bees so some of the swarm queens (which are unmated queens) leave with more swarms. Sometimes a colony has a several afterswarms.

Generally swarming is considered a bad thing because you usually lose those bees. But if you catch them it's a bonus because swarms are notorious for building up quickly. The bees are focused on it already and it's in the natural order of things. Back in the days of skeps and box hives it was always considered a good thing. It was a chance to make increase.

## Causes of swarming

It's good to realize that swarming is the normal response of a hive to success. It means they are doing well enough to reproduce the hive. It is the natural order of things. However, it is inconvenient for the beekeeper to have them swarm, so let's think about what causes them to want to swarm.

First there are two main types of swarms. There are reproductive swarms and there are overcrowding swarms. There are a variety of pressures that push them toward swarming.

## Overcrowding swarm

Since it's the simplest and can happen anytime, let's briefly look at the overcrowding swarm. The factors that seem to contribute are:

No place to put nectar so it gets stored in the brood nest. Prevention: add supers.

Honey or pollen clogging the brood nest so that the queen has nowhere to lay. Prevention: remove combs of honey and add empty frames so that the bees will be occupied drawing wax and the queen will have somewhere to lay and the bees will have more room to cluster in the brood nest.

No place to cluster near the brood nest. The bees like to cluster near the queen (who is in the brood nest) and this clogs the brood nest making it crowded. Prevention: Slatted racks give room to cluster under the brood nest. Follower boards on the outside give room to cluster on the sides of the brood nest. These are a $3/4$" wide top bar with a sheet of plywood or Masonite or similar material in the middle the size of a frame. One on each end replaces one frame in the brood nest.

Too much traffic congesting the brood nest. Prevention: a top entrance will give foragers a way in without going through the brood nest.

So basically, if you keep supers on and provide ventilation you can prevent an overcrowding swarm.

## Reproductive swarm

The bees have been working toward this goal since last winter when they tried to go into winter with enough excess stores to build up in the spring before the flow enough to cause a swarm that will then have the optimum chance to build up enough to survive the following winter.

The first mistake people make about preventing swarms is they think you can just throw on some supers and they won't swarm. But they will. Yes, it's nice to have room for them to store the honey, so the supers are helpful, but the bees intend to swarm and the supers will not deter them from the plan to do a reproductive swarm.

Back to the sequence in the spring, the bees, during winter, rear little spurts of brood. The queen lays a little and they start rearing that batch, but they don't start any new brood until that brood emerges and they take a break. Then they rear another little batch. When pollen starts coming in they start to rear more brood to build up. They also start using up the honey they have stored. This is used to feed brood and also it makes room for more brood.

When the bees think they have enough bees they start filling all of that back in with honey, both to stop the queen from laying, and to have adequate stores in case the main flow doesn't pan out. As the brood nest gets backfilled it makes more and more unemployed nurse bees. These nurse bees start doing a keening buzz that is quite different from the typical harmonious buzz you usually hear—more of a warble. Once the brood nest is mostly full of honey they start swarm cells. About the time they get capped the old queen leaves with a large number of bees. Even if you catch the swarm, the hive has still stopped brood production and has lost (to the swarm) a lot of bees. It's doubtful it will make honey. If there are still enough bees, the hive will throw afterswarms with virgin queens heading them.

If I don't catch them in time, once they make up their mind I always make splits because not much will dissuade them. Destroying queen cells only postpones the inevitable, at best and most likely will leave them queenless. My guess is that most people destroy the queen cells *after* the hive has swarmed without realizing it.

If you catch them trying to swarm between about two weeks and just before the main flow, a cut down split with the old queen and all but one frame of the open brood in a new location is a nice swarm prevention method. Leave the old hive with all the capped brood, one frame of eggs/open brood, no queen and empty supers. Usually, the old hive won't swarm because they have no queen and hardly any open brood. Usually the new hive won't swarm because they have no foragers. This is best done just before the main honey flow.

I often just put every frame that has some queen cells on it with a frame of honey in a two frame nuc to get good queens.

But, of course, the real object is to avoid the swarm and the split (unless you want to do the cut down split) so you'll have a bigger stronger hive that will make more honey.

## Preventing swarming

I do love to catch swarms but who has time to watch the hives all the time to catch them? And if you have that much time, then you have the time to prevent them.

## Opening the broodnest

This, of course is what we want to do. What we need to do is interrupt the chain of events. The easiest way is to keep the brood nest open. If you keep the brood nest from backfilling and if you occupy all those unemployed nurse bees then you can change their mind. If you catch it before they start queen cells, you can put some empty frames in the brood nest. Yes, empty. No foundation. Nothing. Just an empty frame. Just one here and there with two frames of brood between. In other words, you can do something like: BBEBBEBBEB where B is brood comb and E is an empty frame. How many you insert depends on how strong the cluster is. They have to fill all those gaps with bees. The gaps fill with the unemployed nurse bees who begin festooning and building comb. The queen will find the new comb and about the time they get about $1/4$" deep, the queen will lay in them. You have now "opened up the brood nest". In one step you have occupied the bees that were preparing to swarm with wax production followed by nursing, you've expanded the brood nest, and you've given the queen a place to lay. If you don't have room to put the empty combs in, then add another brood box and move some brood combs up to that box to make the room to add some to the brood nest. In other words, then the top box would probably be something like EEEBBBEEEE and the bottom one

BBEBBEBBEB. The other upside is I get good natural sized brood comb.

A hive that doesn't swarm will produce a *lot* more honey than a hive that swarms.

## Checkerboarding aka Nectar Management

Checkerboarding is a technique originated by Walt Wright that involves interspersing drawn and capped honey *over* the brood nest. It in no way involves the brood nest itself. If you'd like to know about this technique and a *lot* more detail about swarm preparation and what goes on in a hive at any given time in the buildup, I would contact Walt Wright. This is a method that also fools the bees into believing that the time has not yet come to swarm. It works without disturbing the brood nest. Basically it's putting alternating frames of empty drawn comb and capped honey directly *above* the brood nest. If you would like to purchase a copy of Walt's manuscript, it's about 60 pages long and last I heard was $8 in a pdf by email or $10 on paper. You can contact him at this address: Walt Wright; Box 10; Elkton, TN 38455-0010(WaltWright@hotmail.com).

# Splits

### *What is the desired outcome?*

I would choose my method for doing a split depending on what you want for an outcome.

Reasons for doing a split:

- To get more hives.
- To requeen.
- To get more production.
- To get less production (for people who don't want too many hives or too many bees).
- To raise queens.
- To prevent swarms.

### *Timing for doing a split:*

As soon as commercial queens are available, or as soon as drones are flying, depending on if you want to buy or raise queens, you *can* do a split. It depends again on what you want for an outcome.

There are an infinite variety of methods for doing a split. Many of these are because of the desired outcome (swarm prevention, maximizing yields, maximizing bees etc.) Some of the variations are also due to buying queens or letting the bees raise queens.

The simple version is to make sure you have some eggs in each of the deeps and put them facing toward the old location. In other words put a bottom board on the left facing the left side of the hive and one on the right facing the right side of the hive and put one deep on each and maybe an empty deep on top of that. Put the tops on and walk away.

There are an infinite number of variations of this.

***The concepts of splits are:***

• You have to make sure that both of the resulting colonies have a queen or the resources to make one (eggs or larvae that just hatched from the egg, drones flying, pollen and honey, plenty of nurse bees).
• You have to make sure that both of the resulting colonies get an adequate supply of honey and pollen to feed the brood and themselves.
• You have to make sure that you account for drift back to the original site and insure that both resulting colonies have enough population of bees to care for the brood and the hive they have.
• You need to respect the natural structure of the brood nest. In other words, brood combs belong together. Drone brood goes on the outside edge of the brood and pollen and honey go outside that.
• You need to allow enough time at the end of the season for them to build up for winter in your location.
• The old adage is that you can try to raise more bees or more honey. If you want both, then you can try to maximize honey in the old location and bees in the new split. Otherwise most splits are either a small nuc made up from just enough to get it started, or an even split.
• The size greatly impacts how quickly it builds from there. You can make a split as small as a frame of brood and a frame of honey. But you can't expect that to raise a well nourished queen. You also can't expect that nuc to build up to a hive by winter. But it makes a good mating nuc or a good place to hold a queen for a while. On the other hand you can make a split that is a minimum of 10 deep frames of bees and brood and honey or 16 medium frames of bees and brood and honey and it will build up rapidly because it has enough "income" and workers to cover its overhead and make a

good "profit". They are at "critical mass" and can really grow rapidly rather than struggle to get by. It's more productive and will build up more quickly to do a strong split, let both double in size and do another strong split than to do four weak splits and wait for them to build up.

**Kinds of splits**

**An even split**

You take half of everything and divide it up. That's an even split. I would face both of new hives at the sides of the old hive so the returning bees aren't sure which one to come back to. In a week or so, swap places to equalize the drift to the one with the queen.

**A walk away split**

Mostly this refers to not giving them a queen and just doing a split by whatever method and walking away and letting them sort things out. Come back in four weeks and see if the queen is laying. But it could also be an even split.

**Swarm control split**

Ideally you want to prevent swarming and not have to split. But if there are queen cells I usually put every frame with any queen cells in its own nuc with a frame of honey and let them rear a queen. This usually relieves the pressure to swarm and gives me very nice queens. But even better, put the old queen in a nuc with a frame of brood and a frame of honey and leave one frame with queen cells at the old hive to simulate a swarm. Many bees are now gone and so is the old queen. Some people do the other kinds of splits (even walk away etc.) in order to prevent swarming. I think it's better to just keep the brood nest open.

## A cut down split

### Concepts of a cut down:

The concepts of a cut down are that you free up bees to forage because they have no brood to care for, and you crowd the bees up into the supers to maximize them drawing comb and foraging. This is especially useful for comb honey production and more so for cassette comb honey production, but will produce more honey regardless of the kind of honey you wish to produce.

This is very timing critical. It should be done shortly before the main honey flow. Two weeks before, would be ideal. The purpose is to maximize the foraging population while minimizing swarming and crowding the bees into the supers. There are variations on this, but basically the idea is to put almost all the open brood, honey and pollen and the queen in a new hive while leaving all the capped brood, some of the honey and a frame of eggs with the old hive with less brood boxes and more supers. The new hive won't swarm because it doesn't have a workforce (which all returns to the old hive). The old hive won't swarm because it doesn't have a queen or any open brood. It will take at least six weeks or more for them to raise a queen and get a decent brood nest going. Meantime, you still get a lot of production (probably a lot *more* production) from the old hive because they are not busy caring for brood. You get the old hive requeened and you get a split. Another variation is to leave the queen with the old hive and take *all* the open brood out. They won't swarm right away because the open brood is gone. But I think it's riskier as far as swarming to crowd a hive with a queen.

## Confining the queen

Another variation on this is to just confine the queen two weeks before the flow so there is less brood to care for and free up nurse bees to forage. This also helps with Varroa as it skips a brood cycle or two. This is a good choice if you don't want more hives and you like the queen. You can put her in a regular cage or put her in a #5 hardware cloth push in cage to limit where she can lay. They will eventually chew under the hardware cloth cage, but it should set her back for a while.

## Cut down Split/Combine

This is a way to get the same number of hives, new queens and a good crop. You set up two hives right next to each other (touching would be good) early in the spring. Two weeks before the main flow you remove all the open brood and most of the stores from both hives, and the queen from one hive, and put it in a hive at a different location (the same yard is fine, but a different place). Then you combine all the capped brood, the other queen, or a new queen (caged), or no queen and one frame with some eggs and open brood (so they will raise a new one) into one hive in the middle of the old locations so all the returning field bees come back to the one hive.

***Frequently Asked Questions about splits***

## How early can I do a split?

It's very difficult for a split to build up unless it has an adequate number of bees to keep the brood warm and reach critical mass of workers to handle the overhead of a hive. For deeps this is usually ten deep frames of bees with six of them brood and four of them honey/pollen in each part of the split. For mediums this is usually sixteen medium frames of bees with ten of them brood and six of them honey/pollen. I'd say you can split as early as you can put together nucs that are this strong. Half this size can work but a stronger split will take off better. Later in the year when it's not frosting occasionally at night, you could get by with somewhat less, but you'll still do better with this much.

## How many times can I split?

Some hives you can't do any splits as they are struggling and never get on their feet. Some hives are such boomers that you can do five splits in a year, although you probably won't get a honey crop.

The object shouldn't be how many can you make, but to keep all the splits you make at critical mass. Critical mass is that point where they are no longer living hand to mouth and they have enough stores, workers, nurse bees and brood to have a surplus. Think of it as economics. If you have barely enough money to pay your bills (or even fall behind on them) you are struggling. When you get to the point where you can pay your bills, you can start to get ahead. When you get to the point where you have some money in the bank and you have a surplus of cash, then life gets pretty easy. Prosperity tends to lead to more prosperity as you can now do things right instead of just getting by. Try it another way. If you run a store you're not really coming out ahead until you cover your overhead.

A hive needs a certain amount of workers to feed the brood (it takes a lot of nurse bees to keep up with a prolific queen), haul the water, pollen, propolis and nectar to feed the brood, build the comb, guard the nest from ants and hive beetles, guard the entrance from skunks and mice and hornets etc.

Once that overhead has been met they can start working on a surplus. If your splits are strong enough to meet their overhead they can take off quickly. If they have barely the resources and workers to survive, they will struggle and take a long time to start really building up.

If you make strong splits and you don't weaken your hives too much you have a shot at getting more splits because they grow faster and more efficiently. Also if you don't weaken your main hives you have more surplus bees to make a surplus crop.

If you take only a frame of brood from each of your strong hives every week they will tend to just make up the difference very quickly with hardly a noticeable lull. One frame of brood and one of honey from each hive put together to fill a ten frame box has a good chance of taking off quickly as opposed to only a few frames of bees.

## How late can I do a split?

What you really need to ask yourself is "when is the best time to do a split". By the bee's example that would be sometime before the main flow so they have a flow to get established on. However this tends to cut into your harvest, so you could do them right after the main flow and probably still have time to build up for the fall, if you make them strong enough and give them a mated queen. Of course this depends on the typical flow where you are. If you typically have a dearth after the flow, you may have to feed if you do this

I'm in Greenwood, Nebraska. In a year with a good fall flow, I can do a split on the 1st of August that may build up enough to overwinter in one or two eight frame medium boxes. But if the fall flow fails they may not build up at all.

**How far?**

The question often seems to come up, how far away to put the split. Mine are usually touching. You need to account for drift if it is less than 2 miles. I've been beekeeping since about 1974 and I've never taken a split 2 miles away unless that's where I wanted to take them anyway. I just do the split and shake in some extra bees or do the split and face both hives to the old location. In other words where the old hive was is where both of the new hives face. Returning bees have to choose. Sometimes I swap those after a few days if one is a lot stronger Usually the one with the queen is stronger.

I say all of this, mostly because it's "the right thing to do", but really since I went to eight frame mediums and since I expanded to 200 hives, I just split by the box and I do nothing about drift. I put two bottom boards where ever there is room on the stand and "deal" the boxes like cards. "One for you and one for you". I add as much empty room as I have boxes full of bees (in other words I double their actual space). So if there are three boxes full of bees on each stand I add three empty supers with frames. But these are strong splits from booming hives with at least two eight frame medium boxes full of bees in each resulting hive.

# Natural Cell Size

**And its implications to beekeeping and Varroa mites**

> *"Everything works if you let it"*—
> Rick Nielsen of Cheap Trick

There has been much talked about and written about small cell and natural cell in recent times and the relationship of small cell to Varroa. Let's clarify a few points about natural cell size.

**Does Small Cell = Natural Cell?**

Small cell has been purported by some to help control Varroa mites. Small cell is 4.9 mm cell size. Standard foundation is 5.4 mm cell size. What is natural cell size?

**Baudoux 1893**

Made bees larger by using larger cells. Pinchot, Gontarski and others got the size up as large as 5.74 mm. But Al Root's first foundation was 5 cells to an inch which is 5.08 mm. Later he started making it 4.83 cells per inch. This is equivalent to 5.26 mm. (*The ABC and XYZ of Bee Culture*, 1945 edition, pages 125-126.)

**Eric Sevareid's Law**

> *"The chief cause of problems is solutions."*

## Foundation Today

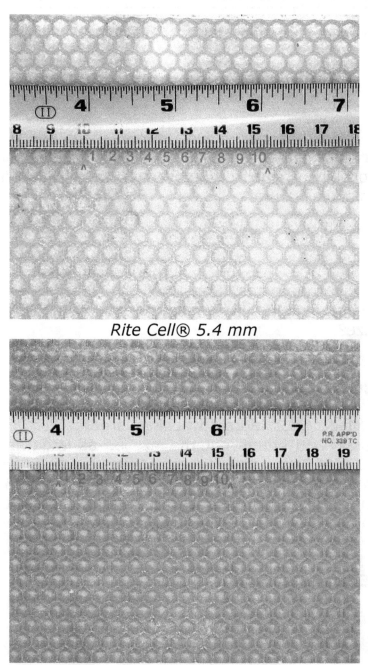

*Rite Cell® 5.4 mm*

*Dadant normal brood 5.4 mm*

# The Practical Beekeeper

*Pierco Medium Sheet 5.2mm*

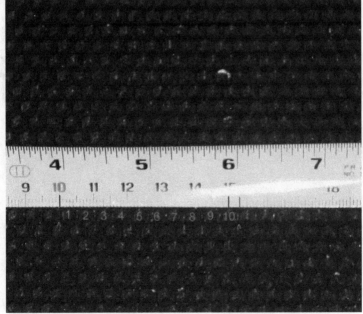

*Pierco Deep Frame 5.25mm*

# Natural Cell Size

Mann Lake PF120 Medium frame

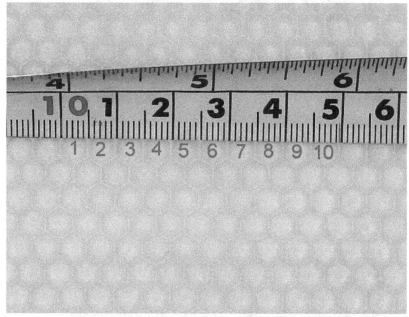

Mann Lake PF100? Deep? frame
NOTE: The Mann Lake PF100 and PF120 are not the same cell size as Mann Lake PF500 and PF520 frames which are 5.4mm.

# The Practical Beekeeper

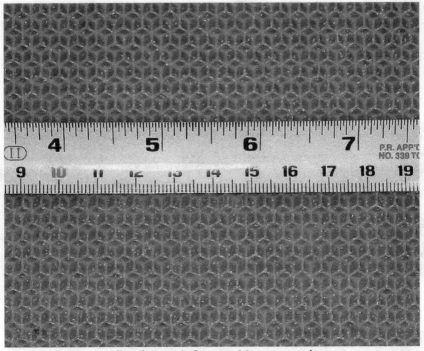

*Dadant 4.9mm Measured*

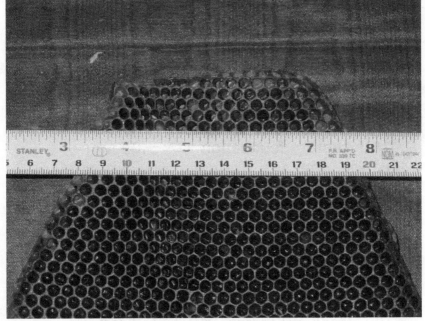

*4.7mm natural comb*

*4.7mm Comb Measurement*

---

**Chart of Cell Sizes**

| | |
|---|---|
| Natural worker comb | 4.6 mm to 5.1 mm |
| Lusby | 4.83mm average |
| Dadant 4.9mm Small Cell | 4.9 mm |
| Honey Super Cell | 4.9 mm |
| Wax dipped PermaComb | 4.9 mm |
| Mann Lake PF100 & PF120 | 4.95 mm |
| 19th century foundation | 5.05 mm |
| PermaComb | 5.05 mm |
| Dadant 5.1mm Small Cell | 5.1 mm |
| Pierco foundation | 5.2 mm |
| Pierco deep frames | 5.25 mm |
| Pierco medium frames | 5.35 mm |
| RiteCell | 5.4 mm |
| Standard worker foundation | 5.4 to 5.5mm |
| 7/11 | 5.6 mm |
| HSC Medium Frames | 6.0 mm |

Drone                           6.4 to 6.6 mm

Note: fully drawn plastic (PermaComb and Honey Super Cell) is always .1mm larger at the mouth than the bottom and you have to allow for the thicker cell wall to come up with an equivalent. So the actual equivalent is pretty much the inside diameter of the mouth.

---

What I've done to get natural comb
- Top Bar Hive
- Foundationless Frames
- Blank Starter Strips
- Free Form Comb
- Empty Frame Between Drawn Combs

---

How much difference between natural and "normal"? Keep in mind that "normal" foundation is 5.4 mm and natural cell is between 4.6 mm and 5.0 mm.

**Volume of cells**

According to Baudoux:

| Cell Width | Cell Volume |
|---|---|
| 5.555 mm | 301 mm$^3$ |
| 5.375 mm | 277 mm$^3$ |
| 5.210 mm | 256 mm$^3$ |
| 5.060 mm | 237 mm$^3$ |
| 4.925 mm | 222 mm$^3$ |
| 4.805 mm | 206 mm$^3$ |
| 4.700 mm | 192 mm$^3$ |

From *The ABC and XYZ of Bee Culture*, 1945 edition, p. 126.

***Things that affect cell size***

- Worker intention for the comb at the time it was drawn:
    - Drone brood
    - Worker brood
    - Honey storage
- The size of the bees drawing the comb
- The spacing of the top bars

---

***What is Regression?***

Large bees, from large cells, cannot build natural sized cells. They build something in between. Most will build 5.1 mm worker brood cells.

The next brood cycle will build cells in the 4.9 mm range.

The only complication with converting back to natural or small cell is this need for regression.

---

***How do I regress them?***

To regress, cull out empty brood combs and let bees build what they want (or give them 4.9 mm foundation)

After they have raised brood on that, repeat the process. Keep culling out the larger combs.

How do you cull out the larger combs? Keep in mind it is normal procedure to steal honey from the bees. It is frames of brood that are our issue. The bees try to keep the brood nest together and have a maximum size in mind. If you keep feeding in empty frames in the center of the brood nest, put them between straight combs to get straight combs, they will fill these with comb and eggs. As they fill, you can add another frame. The brood nest expands because you keep spreading it out to put in the frames. When the large cell frames are too far from the center (usually the outside wall) or when they are contracting the brood nest in the Fall, they will fill them with honey after the brood emerges and then you can harvest them. You could also move the capped large cell brood above an excluder and wait for the bees to emerge and then pull the frame.

**Please** do not confuse this issue of regression. I seem to get questions constantly asking whether to install a package on 5.4mm foundation first since they can't draw 4.9mm foundation well. If you want to get back to natural or small cell size, it is **never** to your advantage to use the already too large foundation they are already using. That is simply going nowhere at all. With a package, if you do so, you will have missed the opportunity to get a full step of regression. Dee Lusby's method is to do shakedowns (shake all the bees off of all the combs) onto 4.9mm foundation and then another shakedown onto 4.9mm to finish the main regression and then cull out the large comb until they have all 4.9mm in the brood nest. Shakedowns are the fastest method but also a stressful method and when you buy a package you already *have* a shakedown. I would take advantage of it. If you intend to get back to natural size then *stop* using large cell foundation all together. The main challenge is getting all the large cell comb *out* of the hive, so don't make that harder by putting more *in*.

Another misconception seems to be that there are large losses in regressing. Dee Lusby went cold turkey, no treatments and only did shakedowns. She lost a lot of bees in the process. Many who tried the same also did. But this is not necessary.

First of all, there is no stress in letting them build their own comb. It's what they have always done. Second, it's not necessary to do shakedowns, it's just quicker. Third, you don't have to go cold turkey on treatments. You can monitor mites (and I would) until things are stable. Meanwhile you could use some non-contaminating treatment *if* the numbers get too high. I have seen no losses from Varroa from regressing in this manner and no increase in losses to stress related problems and I found no need for any treatments.

### *Observations on Natural Cell Size*

First there is no one size of cells nor one size of worker brood cells in a hive. Huber's observations on bigger drones from bigger cells was directly because of this and led to his experiments on cell size. Unfortunately, since he couldn't get foundation at all, let alone different sizes, these experiments only involved putting worker eggs in drone cells which, of course, failed. The bees draw a variety of cell sizes which create a variety of bee sizes. Perhaps these different subcastes serve the purposes of the hive with more diversity of abilities

The first "turnover" of bees from a typical hive (artificially enlarged bees) usually builds about 5.1 mm cells for worker brood. This varies a lot, but typically this is the center of the brood nest. Some bees will go smaller faster.

The next generation of bees, given the opportunity to draw comb will build worker brood comb in the range of 4.9 mm to 5.1 mm with some smaller and some larger. The spacing, if left to these "regressed" bees is typically 32 mm or $1^1/_4$" in the center of the brood nest. Subsequent generations may go slightly smaller.

## Observations on Natural Frame Spacing

### $1^1/_4''$ spacing agrees with Huber's observations

> *"The leaf or book hive consists of twelve vertical frames... and their breadth fifteen lines (one line= $^1/_{12}$ of an inch. 15 lines = $1^1/_4''$). It is necessary that this last measure should be accurate." François Huber 1789*

### Comb Width (thickness) by Cell Size

According to Baudoux (note this is the thickness of the comb itself and not the spacing of the comb on centers)

| Cell Size | Comb width |
|---|---|
| 5.555 mm | 22.60 mm |
| 5.375 mm | 22.20 mm |
| 5.210 mm | 21.80 mm |
| 5.060 mm | 21.40 mm |
| 4.925 mm | 21.00 mm |
| 4.805 mm | 20.60 mm |
| 4.700 mm | 20.20 mm |

*The ABC and XYZ of Bee Culture* 1945 edition Pg 126

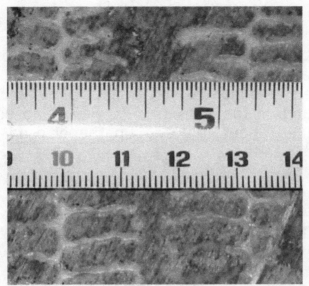

*Wild Comb in Top Feeder Comb Spacing Comb Spacing 30mm*

Here is a brood nest that moved into a top feeder even with plenty of room in the boxes and the inner cover after removing the comb. Spacing on naturally drawn brood comb is sometimes as small as 30 mm but typically 32 mm.

**Pre and Post Capping Times and Varroa**

8 hours shorter capping time halves the number of Varroa infesting a brood cell.

8 hours shorter post-capping time halves the number of offspring of a Varroa in the brood cell.

Accepted days for capping and post capping (based on observing bees on 5.4 mm comb):
Capped 9 days after egg laid
Emerges 21 days after egg laid

## Huber's Observations

Huber's *Observations on Capping and Emergence on Natural Comb.*

Keep in mind that on the 1st day no time has elapsed and on the 20th 19 days have elapsed. If you have doubts about this add up the elapsed time he refers to. It adds up to $18^1/_2$ days.

> *"The worm of workers passes three days in the egg, five in the vermicular state, and then the bees close up its cell with a wax covering. The worm now begins spinning its cocoon, in which operation thirty-six hours are consumed. In three days, it changes to a nymph, and passes six days in this form. It is only on the twentieth day of its existence, counting from the moment the egg is laid, that it attains the fly state."*—François Huber 4 September 1791.

## My Observations

My Observations on Capping and Emergence on 4.95mm Comb.

I've observed on commercial Carniolan bees and commercial Italian bees a 24 hour shorter pre capping and 24 hour shorter post capping time on 4.95 mm cells in an observation hive.

My observations on 4.95 mm cell size
  Capped 8 days after laid
  Emerged 19 days after laid

## Why would I want natural sized cells?

Less Varroa because:
- Capping times shorter by 24 hours resulting in less Varroa in the cell when it's capped
- Postcapping times shorter by 24 hours resulting in less Varroa reaching maturity and mating by emergence
- More chewing out of Varroa

## How to get natural sized cells

Top Bar Hives:
Make the bars 32 mm (1 $^1/_4$") for the brood area
Make the bars 38 mm (1 $^1/_2$") for the honey area

Foundationless frames:
Make a "comb guide" like Langstroth did (see *The Hive and the Honey-Bee*)
Also helpful to cut down end bars to 32 mm (1 $^1/_4$") or
Make blank starter strips
Use a brine-soaked board and dip it in wax to make blank sheets. Cut these into $^3/_4$" wide strips and put in the frames.

## How to get small cells

Use 4.9 mm foundation or
Use 4.9 mm starter strips

### So what Are natural sized cells?

I have measured a lot of natural drawn combs. I have seen worker brood in the range of 4.6 mm to 5.1 mm with most in the 4.7 to 4.8 ranges. I have not seen any large areas of 5.4 mm cells. So I would have to say:

### Conclusions:

Based on my measurements of natural worker brood comb:
- There is nothing *un*natural about 4.9 mm worker cells.
- 5.4 mm worker cells are not the norm in a brood nest.
- Small cell and natural cell have been adequate for me to have hives that are stable against Varroa mites with no treatments.

---

### Frequently asked questions:

**Q:** Doesn't it take longer for them to draw their own combs?

**A:** I have not found this to be true. In my observation (and others who have tried it), they seem to draw plastic with the most hesitation, wax with a little less hesitation and their own comb with the most enthusiasm. In my observation, and some others including Jay Smith, the queen also prefers to lay in it.

**Q:** If natural/small cell size will control Varroa, why did all the feral bees die off?

**A:** The problem is that this question typically comes with several assumptions.

The first assumption is that the feral bees have all but died out. I have not found this to be true. I see a lot of feral bees and I see more every year.

The second assumption is that when some of the feral bees did die, that they all died from Varroa mites. A lot of things happened to the bees in this country including Tracheal mites, and viruses. I'm sure some of the survival from some of this is a matter of selection. The ones that couldn't withstand them died.

The third assumption is that huge numbers of mites hitchhiking in on robbers can't overwhelm a hive no matter how well they handle Varroa. Tons of crashing domestic hives were bound to take a toll. Even if you have a fairly small and stable local population of Varroa, a huge influx from outside will overwhelm a hive.

The fourth assumption is that a recently escaped swarm will build small cell. They will build something in between. For many years most of the feral bees were recent escapees. The population of feral bees was kept high by a lot of recent escapees and, in the past, those escapees often survived. It's only recently I've seen a shift in the population to be the dark bees rather than the Italians that look like they are recent. Large bees (bees from 5.4 mm foundation) build an in between sized comb, usually around 5.1 mm. So these recently swarmed domestic bees are not fully regressed and often die in the first year or two.

The fifth assumption is that small cell beekeepers don't believe there is also a genetic component to the survival of bees with Varroa. Obviously there are bees that are more or less hygienic and more or less able to deal with many pests and diseases. Whenever a new disease or pest comes along the ferals have to survive them without any help.

The sixth assumption is that the feral bees suddenly died. The bees have been diminishing for the last 50 years fairly steadily from pesticide misuse, loss of habitat and forage, and more recently from bee paranoia. People hear about AHB and kill any swarm they see. Several states have killing all feral bees as their official policy.

**Q:** If bees are naturally smaller why didn't anyone notice? Also why are the bee scientist saying they are larger?

**A:** I don't know why, they are saying they are larger, perhaps some of it comes back to the regression issue. If you take bees from large cell comb and let them build what they want, what will they build? Is this the same as natural comb? Sometimes we just have differences in observations because of a variety of factors being involved.

I really don't think it should be hard to accept that they are naturally smaller since there have been plenty of measurements taken over the centuries. Dee Lusby's writings (available on www.beesource.com have references to many articles and discussions on the size of bees and comb and the concept of enlarging it. We have plenty of easy to find evidence that bees used to be smaller.

Find *ABC & XYZ of Bee Culture* books and look under "Cell Size".

Here are some quotes from them.

*ABC & XYZ of Bee Culture*, 38th edition (1980), page 134:

*"If the average beekeeper were asked how many cells, worker and drone comb, there were to the inch, he would undoubtedly answer five and four, respectively. Indeed some text books on bees carry that ratio. Approximately it is correct, enough for the bees, particularly the queen. The dimensions must be exact or there is a protest. In 1876 when A.I. Root, the original author of this book, built his first roll comb foundation mill, he had the die faces cut for five worker cells to the inch. While the bees built beautiful combs from this foundation, and the queen laid in the cells, yet, if given a chance they appeared to prefer their own natural comb not built from comb foundation. Suspecting the reason, Mr. Root then began measuring up many pieces of natural comb when he discovered that the initial cells, five to the inch, from his first machine were slightly too small. The result of his measurements of natural comb showed slightly over 19 worker cells to four inches linear measurement, or 4.83 cells to one inch."*

Roughly this same information is in the 1974 version, page 136 and the 1945 version, page 125. The 1877 version, page 147 says:

> *"The best specimens of true worker-comb, generally contain 5 cells within the space of an inch, and therefore this measure has been adopted for the comb foundation."*

All of the following historic references list that same measurement, 5 cells to the inch and can be reviewed at Cornell's Hive and the Honey Bee Collection online (http://bees.library.cornell.edu/):
- *Beekeeping*, Everett Franklin Phillips pg 46
- *Rational Bee-keeping*, Dzierzon pg 8 and again on pg 27
- *British Bee-keeper's Guide Book*, T.W. Cowan pg 11
- *The Hive and the Honey Bee*, L.L. Langstroth pg 74 of the 4th edition but is in all of them

This "5 cells to the inch" in *ABC XYZ* is followed in all but the 1877 version with a section on "will larger cells develop a larger bee" and information on Baudoux's research.

### So let's do the math:

Five cells to an inch, the standard size for foundation in the 1800s and the commonly accepted measurement from that era, is five cells to 25.4mm which is ten cells to 50.8 mm which is, of course, 5.08mm per cell. This is 3.2 mm smaller than standard foundation is now.

A.I. Root's measurement of 4.83 cells to an inch is 5.25 mm which is 1.5 mm smaller than standard foundation. Of course if you measure comb much you'll find a lot of variance in cell size, which makes it very difficult to say exactly what size natural comb is. But I have measured (and photographed) 4.7 mm comb from commercial Carniolans and I have photographs of comb from bees on natural comb in Pennsylvania that are 4.4mm. Typically there is a lot of variance with the core of the brood nest the smallest and the edges the largest. You can find a lot of comb from 4.8 mm to 5.2 mm with most of the 4.8 mm in the center and the 4.9 mm, 5.0 mm and 5.1 mm moving out from there and the 5.2 mm at the very edges of the brood nest.

> *"Until the late 1800s honeybees in Britain and Ireland were raised in brood cells of circa 5.0 mm width. By the 1920s this had increased to circa 5.5 mm."— John B. McMullan and Mark J.F. Brown, The influence of small-cell brood combs on the morphometry of honeybees (Apis mellifera)—John B. McMullan and Mark J.F. Brown*

Hubersaid in Volume two of *Huber's Observations on Bees* (see translation by C.P. Dadant) that worker cells are $2\text{-}^2/_5$ lines which is equal to 5.08mm which is identical to the early ABC XYZ of Bee Culture.

The 41st edition of ABC XYZ of Bee Culture on Page 160 (under Cell Size) says:

> *"The size of naturally constructed cells has been a subject of beekeeper and scientific curiosity since Swammerdam measured them in the 1600s. Numerous subsequent reports from around the world indicate that the diameter of naturally constructed cells ranges from 4.8 to*

> 5.4mm. Cell diameter varies between geographic areas, but the overall range has not changed from the 1600s to the present time."

And further down:

> "reported cell size for Africanized honey bees averages 4.5-5.1mm."

Marla Spivak and Eric Erickson (Do measurements of worker cell size reliably distinguish Africanized from European honey bees (Apis mellifera L.)?*American Bee Journal*, April 1992, p. 252-255 says:

> "...a continuous range of behaviors and cell size measurements was noted between colonies considered "strongly European" and "strongly Africanized". "

> "Due to the high degree of variation within and among feral and managed populations of Africanized bees, it is emphasized that the most effective solution to the Africanized "problem", in areas where Africanized bees have established permanent populations, is to consistently select for the most gentle and productive colonies among the existing honey bee population" —
> Identification and relative success of Africanized and European honey bees in Costa Rica..

In my observation, there is also variation by how you space the frames, or variation on how *they* space the combs. 38 mm (1$^1$/$_2$") will result in larger cells than 35 mm (1$^3$/$_8$") which will be larger than 32 mm (1$^1$/$_4$"). In naturally spaced comb the bees will sometimes crowd the combs down to 30 mm in places with 32 mm more common in just brood comb and 35mm more common where there is drone on the comb.

So what is natural comb spacing? It is the same problem as saying what natural cell size is. It depends.

But in my observation, if you let them do what they want, for a couple of comb turnovers, you can find out what the range of these is and what the norm is. The norm was (and is) *not* the standard foundation size of 5.4 mm cells and it is *not* the standard comb spacing of 35 mm.

# Ways to get smaller cells

## How to get natural sized cells

### Top Bar Hives

Make the bars 32 mm ($1^1/_4$") for the brood area
Make the bars 38 mm ($1^1/_2$") for the honey area

### Foundationless frames

Make a "comb guide" like Langstroth did (see Langstroth's *The Hive and the Honey-Bee*)
Also helpful to cut down end bars to 32 mm ($1^1/_4$") or

### Make blank starter strips

Use a brine soaked board and dip it in wax to make blank sheets. Cut these into $^3/_4$" wide strips and put in the frames.

---

### How to get small cells

Use 4.9 mm wax foundation or
Honey Super Cell (see www.honeysupercell.com)
PermaComb or PermaPlus (5.0mm cell size)
Mann Lake PF100s or PF120s (4.95mm cell size).

# Rationalizations on Small Cell Success

This chapter is *not* to talk about my theories of why small cell works or others who are doing it, but the theories of those who want to explain away the success of small cell beekeepers with theories that are more in line with their model of the world. There seem to be many theories from those who are not doing small cell and who want to explain the success of small cell beekeepers in some other frame of reference that makes sense to them. I will address a few of these here.

### AHB

One explanation, which is consistent with other beliefs held by these individuals is that small cell beekeepers must have Africanized honey bees. Since they believe that AHB build smaller cells and EHB do not, in their model of the world, that explains both the size of the cells, and the success with Varroa as well as early emergence and other issues to do with Varroa. The problem with this theory is that many of us are keeping bees in Northern climates, where we are told AHB can't survive, are selling them to others, who comment on how gentle our bees are, have them regularly inspected, without any complaints of aggressiveness or suspicions of AHB from inspectors, and indeed most of us are collecting local survivor stock when we can, which supposedly could not survive in the North if it was AHB. And I have had samples tested at the request of someone doing a study on bee genetics which says they are not. The fact is at least those of us not in AHB areas are definitely not raising AHB and don't want to. Whether or not Dee Lusby, or others in AHB areas end up with some AHB genes, is a different discussion, but it's irrelevant to the fact that most of us small cell beekeepers do not live in AHB areas and are not raising AHB and are not interested in raising AHB yet our bees are surviving.

**Survivor stock**

While it's true that many small cell and natural cell beekeepers try to breed from survivors, this is simply the logical thing to do. You raise bees that can survive where you are. Many people are doing that even if they are not doing small cell and even if it's not for Varroa issues, but just wintering issues. Typically the people using this argument quote the losses that the Lusby's had while regressing as evidence that they just bred stock that could survive the Varroa. This seems plausible if the Lusbys were the only example, but I had no large losses while regressing and started with commercial stock and when I did the same thing on large cell, I lost all of them to Varroa several times over. Starting again with new commercial stock on small cell I have lost none to Varroa. Considering how many people are working so diligently to try to breed resistant stock, I think it's beyond believability that so many of us small cell beekeepers just blundered into Varroa-resistant stock with so little effort. If these people really believe genetics is the cause of our success then they should be begging us to sell them breeder queens. Since they are not, I do not think even they believe this. I certainly don't believe this, although I would love to. It would greatly increase the value of my queens. Since I regressed and since my Varroa issues went away, I then did start breeding from survivor stock I could find around, because I want bees acclimatized to my environment. I have better wintering when I do this. I did not see any change in Varroa issues when doing this as Varroa problems had already disappeared.

## Blind faith

This isn't so much a reason being given that it works, as much as discounting that it does work and trying to find a reason people *think* it works. It seems that a lot of detractors of small cell think that the whole group of small cell beekeepers are fanatically religious followers of Dee Lusby suffering from mass hysteria. The implication is that we are deluded into believing it is working when it is not. Anyone who comes to one of the many organic meetings where Dee Lusby, Dean Stiglitz, Ramona Herboldsheimer, Sam Comfort, Erik Osterlund, I and others speak would see the absurdity of this. As would anyone who participates in the organic beekeepers Yahoo group. We often have different observations and often disagree, as all honest beekeepers do. If we all spouted some standard party line, then this might be a legitimate concern, but while we agree on the basic concepts, we often disagree on details and we have all had different experiences probably caused our locations and our climate as well as just chance. While I have great respect for all of the above listed speakers and particularly for Dee, as she and her late husband Ed pioneered this work, I have never been in total agreement with her or the rest.

The four things I think we all agree on are: No treatments, natural or small sized cells, local adapted stock, and avoiding artificial feed. But while Sam and I are pretty happy with simple foundationless, Dee is more focused on actual specific cell size. While Dee will feed barrels of honey to her bees, I have neither the time nor the honey for such things and will, if they are faced with not enough honey for winter stores, feed sugar. While Dean and Ramona like natural comb, their experience has been that they had to force the bees down with some Honey Super Cell first to get them regressed, while I've often had good luck with just foundationless regressing quickly. This may be related to the genetics or the cell size in the hives that are the source of my packages and their packages. It is difficult to say. The point is, there is no "party line".

### Resistance

Personally, I have never been able to figure out the resistance to the concept of small cell or natural comb. While the large cell beekeepers are obsessed with Varroa, I get to just keep bees. While the large cell beekeepers are still searching for a solution to Varroa, I get to work on my queen rearing and finding easier ways to do less work. Since letting the bees build comb is easier than using foundation, and since those of us doing that are not having Varroa issues, I would think there would be a lot more interest in doing the same. The battle cry of the detractors, of course, is either that there is no study to prove it works, or that there are studies that show that it doesn't. All of this is, of course, irrelevant to me since I'm still not having Varroa issues anymore. I've been hearing such arguments about things not being proved scientifically all my life and yet have lived to see many of those things proved eventually. In the end it's about what works, not what has been proven. In the end it's not about mite counts, although mine have dropped to almost none over time, it's about survival. No one seems to want to count living hives instead of mites, but it's a much easier thing to count and much more meaningful. If you put one beeyard on small cell and leave another on large cell, then it seems like the "last man standing" would be an easy way to decide. If one yard dies out and the other does well, that would seem a much better way to decide than counting mites.

## Small Cell Studies

There are a few positive small cell studies, but also several that show higher mite counts on small cell and people always ask why. I don't know for sure, as it is inconsistent with my experience, but let's look at that. Let's assume a short term study (which all of them have been) during the drone rearing time of the year (which all of them have been) and make the assumption for the moment that Dee Lusby's "pseudodrone" theory is true, meaning that with large cell the Varroa often mistake large cell workers for drone cells and therefore infest them more. The Varroa in the large cell hives during that time would be less successful at reproducing, but cause more damage, because they are in the wrong cells (workers). The Varroa, during that time would be more successful at reproducing but cause less damage to the workers on the small cell because they are in the drone cells. But later in the year this may shift dramatically when, first of all the small cell workers have not taken damage from the Varroa and second of all the drone rearing drops off and the mites, looking for drone cells (or "pseudodrone" cells) have nowhere to go.

In the end, as Dann Purvis says, "It's not about mite counts. It's about survival". No one seems interested in measuring that. What I do know is that after a couple of years the mite counts dropped to almost nothing on small cell. But that did not take place in the first three months.

# Foundationless

## Why would one want to go foundationless?

How about no chemical contamination of the combs and natural Varroa control from natural cell size? As far as contamination, some of my queens are three years old and laying well. I don't think you'll find anyone who is using chemicals in their hives with that kind of longevity and health in their queens. You can also get clean wax combs with natural cells in a top bar hive.

Comb On Blank Starter Strip. The cells are 4.5mm. The frames were spaced 1 $1/4$"

### How do you go foundationless?

Bees need some kind of guide to get them to draw straight comb. Any beekeeper has seen them skip the foundation and build combs between or out from the face of the comb, so we know that sometimes they ignore those clues. But a simple clue like a beveled top bar or a strip of wax or wood or even a drawn comb on each side of an empty frame will work most of the time. You can just break out the wedge on a top bar, turn it sideways and glue and nail it on to make a guide. Or put Popsicle sticks (jumbo craft sticks) or paint sticks in the groove. Or just cut out the old comb in a drawn wax comb and leave a row at the top or all the way around.

*Foundationless Frame*

I made these by ordering frames from Walter T. Kelley with no grooves in the top and bottom bars and cutting the top bars at a 45 degree angle on both sides. Kelley is now offering them already made with the bevel on them. The bees tend to follow the sloped top bar.

*Foundationless Frame*

*Drawn Foundationless Frame*

Note the picture of *Drawn Foundationless Frame.* You can see the corners are often open, the bottom seems to be the last to get attached, but this is attached on all four sides and ready to be uncapped and extracted.

*Dadant Deep Foundationless Frame (11$^{1}/_{4}$ in.)*

Here is a Dadant Deep foundationless with a comb guide all the way around and a $^{1}/_{16}$" steel rod for support horizontally in the center. This allows cutting six pieces of 4" by 4" comb honey out without fighting with wires. Langstroth also used the comb guides on the side like this.

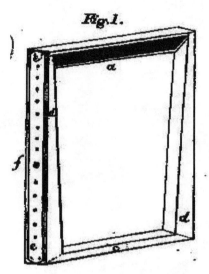

*Beveled Top Bar Frame*

*Langstroth's Foundationless Frame*

L.L. Langstroth has pictures of this design in the original "Langstroth's Hive and the Honey Bee" which you can still buy as a reprint.

**Foundationless frames**

In my experience the bees will draw their own comb faster than they will draw foundation. I'm not the only one to make the observation that bees are not attracted to foundation.

> "Foundation, even composed of pure beeswax, is not intrinsically attractive to bees. Swarming bees offered the opportunity to cluster on foundation or some branch, show no preference for foundation." —The How-To-Do-It Book of Beekeeping, Richard Taylor

**Historic References**

Most of these can be found online in Cornell's Hive and the Honey Bee collection (http://bees.library.cornell.edu/).

> "HOW TO SECURE STRAIGHT COMBS. "The full advantages of the movable comb principle is only secured by getting all the combs built true within the frames. Upon the first introduction of movable frames, bee-keepers frequently failed in this although much care

and attention were given. Mr. Langstroth, for a time, used for guides strips of comb attached to the under side of the top bar of the frame. This is a very good practice when the comb can be had, as it usually secures the object besides giving the bees a start with worker comb. Next followed the triangular comb guide consisting of a triangular piece of wood tacked to the under side of the top bar, leaving a sharp corner projecting downwards. This is a valuable aid and is now universally adopted." —Facts in Bee Keeping by N.H. King and H.A. King 1864, pg. 97

"If some of the full frames are moved, and empty ones placed between them, as soon as the bees begin to build powerfully, there need be no guide combs on the empty frames, and still the work will be executed with the most beautiful regularity." —The Hive and the Honeybee by Rev. L.L. Langstroth 1853, pg. 227

"Improved Comb Bar.—Mr. Woodbury says that this little contrivance has proved very effectual in securing straight combs when guide combs are not obtainable. The lower angles are rounded off whilst a central rib is added of about $1/8$ of an inch in breadth and depth. This central rib extends to within $1/2$ an inch of each end, where it is removed in order to admit of the bar fitting into the usual notch. All that is necessary to insure the regular formation of combs is, to coat the underneath surface of the central rib with melted wax. Mr. Woodbury further says, "my practice is to use plain bars, whenever guide-combs are attainable, as these can be attached with much greater facility to a plain than to a ribbed bar; but whenever I put in a bar without comb, I always use one of the improved ones. By this method, crooked and irregular combs are altogether unknown in my apiary." Most of our bars are made with the ridge; but should any of our customers prefer the flat ones, we keep a few to supply their requirements"—Alfred Neighbour, The Apiary, or, Bees, Bee Hives, and Bee Culture pg 39

"Top bars have been made by some hive manufacturers from one-fourth-inch to three-eighths-inch strips, strengthened somewhat by a very thin strip placed edgewise on the underside as a comb guide; but such bars are much too light and will sag when filled with honey or with brood and honey..."—Frank Benton, The Honey Bee: A Manual of Instruction in Apiculture pg 42

*"Comb Guide.—Generally a wooden edge, or a strip of comb or fdn., in the top of a frame or box, on which comb is to be built...As the comb guide is 9-16, and the cut in the end bar $^3/_4$ we have 3-16 left for whole wood in the top bar, as at A, and the table should be set, as to leave just this amount of wood uncut. Even if the fdn. is fastened in the frames with melted wax as many do, I would have such a comb guide, because it adds so much to the strength of the frame, and obviates the necessity of having a very heavy top bar. The bees will, in time, build their combs right over such a comb guide, and use the cells above the brood for honey."— A.I. Root, ABC of Bee Culture 1879 edition pg 251*

*"A comb guide proper is a sharp edge or corner in the frame, from which the comb is to depend, the bees usually choosing to follow this edge, rather than diverge to an even surface; portions of comb are sometimes used for the same purpose."— J.S. Harbison, The bee-keeper's directory, footnote at the bottom of page 280 and 281*

## FAQs

### Box of empty frames?

**Q:** You mean I can just put a box of empty frames on the hive?

**A:** No. The bees need some kind of guide.

### What is a guide?

**Q:** What is a comb guide?

**A:** It can be any of several things. You can use an empty frame with nothing added *if* you have a drawn brood comb on each side as the brood comb will act as a guide. You can put popsicle sticks in the groove to make a sort of wooden strip, or cut a piece of wood to make a wood starter strip. You can turn the wedge on end and glue it in. You can cut a triangular piece and put on the bottom of the top bar. You can buy chamfer molding and cut it to fit and put it on the bottom of the top bar. You can cut the top bars on a bevel. You can make a sheet of empty wax and cut it into $3/4$" wide strips and put in the groove of the top bar and wax it in. You can cut strips of regular foundation into $3/4$" wide strips and wax that in the groove or nail it with the cleat. If the frame already had comb in it, you can just leave the top row of cells on the top bar for a guide. Any of these work fine.

### Best guide?

**Q:** Which comb guide do you like best?

**A:** I like most them fine, but I like the durability of the beveled top bar and I think the comb is attached a bit better. Next I'd probably go for the wood strip. Last I'd go for the starter strips as they sometimes get hot and fall out if the bees haven't used them yet. But I also feed empty frames into brood nests all the time as I have a lot of old frames around. Bottom line is, I do whatever is the easiest at the time. The worst comb guide is filling a groove with just a wax bead. Wax in the groove is barely a suggestion and not at all a good guide. You need something that protrudes significantly. $1/4$" is good.

**Extract?**

**Q:** Can I extract them?

**A:** Yes. I extract them all the time. Just make sure they are attached on all four sides and the wax isn't so new that it's still soft, like putty. Once the wax is mature and the comb is attached at least some on all four sides, it extracts fine. Of course you should always be gentle with any wax combs (wired or not) when extracting.

**Wire?**

**Q:** Do I need wire in them?

**A:** I don't use wire but I don't use deeps either.

**Q:** *Can* I use wire in them?

**A:** Sure. The bees will incorporate the wire into the comb. Of course you need the hive level anyway, but this becomes more obvious with wire in the comb.

Wire is probably more useful when doing deeps than mediums. I run all mediums.

## Wax them?

**Q:** Do I need to wax them?

**A:** I find wax to be counterproductive. It is more work, it often falls off, and it is never attached to the bar as well as the bees will attach their own comb. I not only *don't* recommend that you wax them, I recommend that you *not* wax them.

## Whole box?

**Q:** Can I put a whole box of foundationless frames on a hive?

**A:** Assuming we mean frames with comb guides, yes, you can. Usually this works fine. Sometimes because of a lack of a comb to use as a "ladder" to get up to the top bars, the bees start building comb up from the bottom bar. For this reason I prefer to have one frame of drawn comb or a full sheet of foundation in a super being added on. This isn't a problem when installing a package. Another reason for the one comb, though, is it's good insurance at getting the combs in the right direction. Another solution to them trying to build comb up, is to put the empty box under the current box so they can work down.

## Will they mess up?

**Q:** Won't the bees mess it up without foundation?

**A:** Sometimes. But they mess it up sometimes even with wax and even more often with plastic. I've seen no more bad combs doing foundationless than I have using plastic foundation. Some of this appears to be genetic as some hives build good comb even when you do everything wrong. Other hives build messed up comb even when you do everything right and simply repeat the "mistakes" when you remove them.

I said it before, but it bears repeating. The most important thing to grasp with any natural comb hive is that because bees build the next comb parallel to the current one, one good comb leads to another in the same way that one bad comb leads to another. You cannot afford to not be paying attention to how they start off. The most common cause of a mess of comb is leaving the queen cage in as they always start the first comb from that and then the mess begins. I can't believe how many people want to "play it safe" and hang the queen cage. They obviously can't grasp that it is almost a guarantee of failure to get the first comb started right, which without intervention is guaranteed to mean every comb in the hive will be messed up. Once you have a mess the most important thing is to make sure the *last* comb is straight as this is always the guide for the *next* comb. You can't take a "hopeful" view that the bees will get back on track. They will not. You have to put them back on track.

This has nothing to do with wires or no wires. Nothing to do with frames or no frames. It has to do with the last comb being straight.

**Slower?**

**Q:** Won't it set the bees back having to build their own comb?

**A:** In my experience, and many others who have tried it as well, the bees build their own comb much more quickly than they build on foundation. Using foundation sets them back in many ways. First they draw foundation more slowly. Second, the foundation is all contaminated with fluvalinate and coumaphos. Third, unless you're using small cell foundation, you're giving them cells that are larger than they want and giving the advantage to the Varroa.

## Beginners

**Q:** Is it a good idea for a beginner to use foundationless?

**A:** In my opinion it's easier for the beginner who has no habits to adjust to foundationless. It's much more difficult for the seasoned beekeeper to adjust to keeping hives perfectly level, not turning comb flatways, not shaking bees vigorously off of a comb that's still new and not well attached etc. Beginners will often break one comb and learn their lesson. Experienced beekeepers will keep falling back into habits and breaking combs for a while until they finally get it ingrained as a new habit.

## If they mess up?

**Q:** What if they mess it all up?

**A:** It's doubtful but possible that they will. I've seen this occur more often when a box full of frames with wax foundation collapses in the heat. I suppose this seems much more frightening to someone who has never done a cutout. If you've ever cut all the combs out of a wild hive and tied them into frames, then you already know what to do. You cut the wild combs and you put them in an empty frame and use rubber bands or string to hold them in the frame. The bees will take care of the rest. They do this just as often with plastic foundation and it's often more difficult to fix.

## Dimensions

**Q:** If I make my own what dimensions should they be?

**A:** You can make them out of standard frames, but I do prefer them with smaller end bars and slightly smaller top bars. See the *Narrow Frames* chapter.

# Narrow Frames

## Observations on Natural Frame Spacing

### 1 ¹/₄" spacing agrees with Huber's observations

> "The leaf or book hive consists of twelve vertical frames... and their breadth fifteen lines (one line= $^1/_{12}$ of an inch. 15 lines = $1^1/_4$"). It is necessary that this last measure should be accurate." François Huber 1806

Brood nest that moved into a top feeder. Inner cover after removing the comb. Spacing on naturally drawn brood comb is sometimes as small as $1^1/_8$" (30mm) but typically $1^1/_4$" (32mm).

## Comb Width by Cell Size

According to Baudoux (note this is the thickness of the comb itself and not the spacing of the comb on centers)

Cell Size    Comb width
5.555 mm  22.60 mm
5.375 mm  22.20 mm
5.210 mm  21.80 mm
5.060 mm  21.40 mm
4.925 mm  21.00 mm
4.805 mm  20.60 mm
4.700 mm  20.20 mm
ABC XYZ of Bee Culture 1945 edition Pg 126

---

## Historic references to narrower frame spacing

"...are placed the usual distance, so that the frames are $1^9/_{20}$ inch from centre to centre; but if it is desired to prevent the production of drone brood, the ends of every other frame are slipped back as shown at B, and the distance of $1^1/_4$ inch from centre to centre may be maintained."—T.W. Cowan, British beekeeper's Guide Book pg 44

"On measuring the combs in a hive that were regularly made, I found the following result, viz; five worker-combs occupied a space of five

> and a half inches, the space between each being three-eights of an inch, and allowing for the same width on each outer side, equals six and a quarter inches, as the proper diameter of a box in which five worker-combs could be build...The diameter of worker-combs averaged four-fifths of an inch; and that of drone-combs, one and one-eight of an inch."—T.B. Miner, The American Bee Keeper's Manual, pg 325

If you take off the extra $3/8$" on the last one this is $5^7/8$" for five combs divided by five is 1.175" or $1^3/16$" on center for each comb.

> "Frame.—As before mentioned, each stock hive has ten of these frames, each 13 inches long by $7^1/4$ inches high, with a $5/8$ inch projection either back or front. The width both of the bar and frame is $7/8$ of an inch; this is less by $1/4$ of an inch than the bar recommended by the older apiarians. Mr. Woodbury,—whose authority on the modern plans for keeping bees is of great weight,—finds the $7/8$ of an inch bar an improvement, because with them the combs are closer together, and require fewer bees to cover the brood. Then too, in the same space that eight old fashioned bars occupied

*the narrower frames admit of an additional bar, so that, by using these, increased accommodation is afforded for breeding and storing of honey."—* Alfred Neighbour, The Apiary, or, Bees, Bee Hives, and Bee Culture...

*"I have found it to be just that conclusion in theory that experiment proves a fact in practice, viz: with frames $7/8$ of an inch wide, spaced just a bee-space apart, the bees will fill all the cells from top to bottom with brood, provided deeper cells or wider spacing, is used in the storage chamber. This is not guess-work or theory. In experiments covering a term of years. I have found the same results, without variation, in every instance. Such being the fact, what follows? In answer, I will say that the brood is invariably reared in the brood-chamber — the surplus is stored, and at once, where it should be, and no brace-combs are built; and not only this, but the rearing of drones is kept well in hand, excess of swarming is easily prevented, and, in fact, the whole matter of bee-keeping work is reduced to a minimum, all that is required being to start with sheets of comb just $7/8$ of an inch thick, and so spaced that they cannot be built any deeper. I trust that I have made myself un-*

*derstood; I know that if the plan indicated is followed, beekeeping will not only be found an easier pursuit, but speedy progress will be made from now on."—"Which are Better, the Wide or Narrow Frames?" by J.E. Pond, American Bee Journal: Volume 26, Number 9 March 1, 1890 No. 9. Page 141*

Note: 7/8" plus 3/8" (max beespace) makes 1 1/4". 7/8" plus 1/4" (min beespace) makes 1 1/8".

*"But those who have given special attention to the matter, trying both spacing, agree almost uniformly that the right distance is $1\,^3/_8$ or, if anything, a trifle scant, and some use quite successfully $1^1/_4$ inch spacing." — The ABC and XYZ of Bee Culture by Ernest Rob Root Copyright 1917, Pg 669*

*"With so many beginners wanting to know about eleven deep frames in a 10 frame deep Langstroth brood chamber I will have to go into further details. But first this letter from Anchorage, Alaska of all places. For that is as far north as you can keep bees. He writes, I'm a new beekeeper with one season's experience with two hives. A good friend is in the same boat he had read one of your articles on "Squeezing" the*

bees and tried one of his hives that way result a hive full of bees and honey. This year we will have eight hives with eleven frames in the brood chamber."

"If you, too, want to have eleven frames in the brood chamber do this. In assembling your frames besides nails use glue. It' a permanent deal anyway. Be sure your frames are the type with grooved top and bottom bars. After assembling the frames, plane down the end bars on each side so that they are the same width as the top bar. Now drive in the staples. As I mentioned last month make them by cutting paper clips in half. They cost but little and don't split the wood. Drive the staples into the wood until they stick out one quarter inch. The staples should be all on one side. This prevents you from turning the frame around in the brood nest. It's a bad practice and it upsets the arrangement of the brood nest. It is being done, but it leads to chilling of brood and it disturbs the laying cycle of the queen. I am talking to beginners, but even old timers should not commit this bad practice. As for the foundation, if you use molded plastic foundation just snap it into the frame and you are ready to go."— *Charles Koover, Bee Culture, April 1979, From the West Column.*

The standard frame width on Hoffman frames is 1³/₈". That means that from center to center combs are spaced 1³/₈" apart. This makes a comb about 1" thick and a beespace between the combs about ³/₈". This spacing works pretty well as an all around spacing and yet beekeepers usually space the frames in the supers further, like 1¹/₂" or more apart. The 1³/₈" was already a compromise between honey storage, drone brood comb and worker brood comb. Natural worker brood comb being spaced 1¹/₄" while natural drone comb is more like 1³/₈" and honey storage typically is about 1¹/₂" or more (1¹/₄"=32mm, 1³/₈" = 35mm and 1¹/₂"=38mm).

**Spacing frames 1¹/₄" has advantages**

Among them:

- Less drone comb.
- More frames of brood in a box.
- More frames of brood can be covered with bees to keep them warm as the layer of bees is only one bee deep instead of two.
- According to some research back in the 70's in Russia, there was less Nosema.
- It's more natural spacing for smaller cells.
- It incites the bees to build smaller cells. The smaller spacing contributes towards them viewing the comb on it as worker comb.

***Frequent misconceptions:***

- That $1^1/_4''$ (32mm) is only right for Africanized honey bees. I've let European honey bees build their own comb and they space worker brood comb as small as $1^1/_8''$ (30mm) but typically $1^1/_4''$ for the core of the brood nest. Wider at the outside edges when they want drones and even wider when they want to store honey.
- That your frames won't be interchangeable with $1^3/_8''$ frames. I interchange them all the time. Many of the historical references above show that people often spaced them tighter in the center and wider on the outside edges. There is nothing stopping you from putting a $1^3/_8''$ frame in the middle of $1^1/_4''$ frames or vice a versa.
- That it simply doesn't matter. Well, it probably doesn't matter a lot but see the above advantages.

***Ways to get narrow frames***

- Assuming no nails from the outsides of the end bars, you can plane off the end bars of regular frames until they are $1^1/_4''$ wide. If you do this before assembling the frames, you can also cut the top bar down to 1" wide on a table saw.
- You can make or buy frames built from scratch. Either by adjusting the dimensions and building Hoffman frames or by building Killion style frames and simply changing the spacing (see "Honey in the Comb" by Carl Killion or later editions Eugene Killion).
- You can intersperse PermaComb (which has no spacers) with regular Hoffman combs and then space them a little further by hand.

- You can build Koover frames (see old 70's Gleanings in Bee Culture articles or plans on nordykebeefarm.com)

## FAQs

**Q:** Won't the top bars be too close if I plane off the end bars?

**A:** A little, but you can get by with it. It does cramp them down to about $3/16$" between the top bars, but bees can get through a $5/32$" hole. I prefer to have more space but not enough to cut down the top bars on regular frames. I do prefer them enough that I make them smaller when I make frames or order them smaller if I can get someone to make them.

**Q:** Why not put 9 frames in the brood box of a ten frame box? Won't that keep things the same (since I want to run nine in my supers) and give them more space so they don't swarm and I don't roll bees pulling out frames?

**A:** In my experience you'll roll more bees with this arrangement (9 in a 10 frame box) because the surface of the comb will be very uneven due to the thickness of the brood being consistent while the thickness of honey storage varies. This means that frame spaced nine in a ten frame box have an uneven surface. That uneven surface is more likely to catch bees between two protruding parts and roll them than when they are even. It also takes more bees to cover and keep warm the same amount of brood when you have 9 frames instead of 10 or 11.

*"...if the space is insufficient, the bees shorten the cells on the side of one comb, thus rendering that side useless; and if placed more than the usual width, it requires a greater amount of bees to cover the brood, as also to raise the temperature to the proper degree for building comb, Second, when the combs are too widely spaced, the bees while refilling them with stores, lengthen the cells and thus make the comb thick and irregular—the application of the knife is then the only remedy to reduce them to proper thickness."—J.S. Harbison, The Beekeeper's Directory, p. 32*

---

# Yearly Cycles

Beekeeping, like any farming, follows the seasons. It is cyclic in nature and the bigger cycle is the year. Smaller cycles are 21 day worker brood cycles etc. but the big picture for beekeeping is a year.

In my view the beekeeper's year starts, as it does for the bees, in preparing the colony for winter. A colony that has a good footing for surviving the winter and prospering in early spring has a good start on the year.

My view, of course will be colored by my experiences in a cold northern climate. You may need to adjust things for your climate.

**Winter**

From a beekeeper's point of view Winter starts at the first killing frost. From this point on the bees will have no resources coming in. No nectar. No pollen. Before this happens they need to be in pretty good shape. Some winters come early and set in early and there are no other opportunities to prepare.

**Bees**

Basically for winter they need to have a sufficient quantity of bees. Lacking this they should either be babied in some way (difficult at best) or combined with another weak hive to make one strong enough to winter. This will vary by race of bee and by climate. Here with Italians I'd want at least a basketball sized cluster. With Carnis, a soccer ball sized cluster and with ferals something between a soccer ball and a softball.

## Stores

They should have enough food to last the winter. I try to leave them enough, but sometimes with a dearth or a poor fall flow they can end up light. Here, in Greenwood, Nebraska, with Italians you need a hive to weigh about 150 pounds. With ferals that's about a 90 pound hive. A light hive can be fed syrup or you can put sugar on top of newspaper on the top bars to make up the deficit. Some people feed pollen or substitute in the late fall as well. Fall syrup is usually 2:1 (sugar:water)

## Setup for winter

They should have no queen excluder and if they have a bottom entrance they should have a mouse guard. A reduced entrance is helpful to prevent robbing. They need to have some kind of top entrance.

### Spring

Spring for the beekeeper starts at the blooming of the maples. Here where I live that's late February or early March. This is when the bees start rearing brood in earnest. It's important from this point on that the supply of pollen and stores is not interrupted as this can interrupt brood rearing. Pollen patties are a common solution for this. Mix pollen with honey to make a dough and roll between waxed paper to make patties. Or feed it open in an empty hive. Feed 1:1 or 2:1 syrup if they are light on stores. On a warm day do a full inspection and check for eggs and brood. Mark the queenless ones to requeen or combine. Clean off the bottom boards and inspect them for dead Varroa mites. If you're using Walt Wright's Nectar management, it's time to checkerboard. If you're not you need to keep an eye on things to prevent early swarms. When the weather starts staying warm enough, open up the brood nest by putting some empty frames in the middle of the brood nest. If it's a booming hive with lots of bees, two or three frames. If it's a moderate hive, one. If it's a weak hive, leave it alone. Don't add a lot of room as the weather is still chilly and too much space is still a stressful thing. The hive is trying to build up enough to swarm before the main flow. Brood rearing has kicked in. Drone rearing will kick in soon.

## *Summer*

Summer, from a beekeepers point of view, is when swarm season hits or just a few weeks before the flow. The flow is when you start seeing white wax and new comb. This is a time to watch for swarm preparations (backfilling the brood nest) and keep the brood nest open. If swarm preparations have progressed to swarm cells, do splits to get spare queens. Add supers for honey storage. By this point, too much room isn't an issue so pile them on the strong hives. Here this would be mid to late May. If you want to do cut down splits or confine a queen for a better crop or to help with Varroa this would be the time. Two weeks before the main flow would be almost perfect timing.

## *Fall*

Fall, from a beekeepers point of view, is when the main summer flow is over and it's time to harvest the honey. The flowers with darker stronger tasting nectar will be blooming soon—goldenrod, smartweed, asters, sunflowers, partridge peas and chicory. It's a good time to requeen as queens are better mated and more available. Also a good time to rear queens, unless there's a bad drought. Towards the end of fall is when you get them settled in for winter. Put on mouse guards. Remove excluders. Remove empty boxes. Reduce entrances. Equalize stores or feed. In other words we are back to setting up for winter.

# Wintering Bees

I have hesitated to write on wintering bees and so far had resisted the temptation because wintering is so tied to locale. But it is a critical issue and I get questions all the time and so I wish to state what I think on many of the issues. Please read all of this with *locale* in mind. I will try to cover what I do in my locale (Southeast Nebraska) in detail and why I do what I do, but that does not mean it is the best for your locale or that some other methods might not work in other or even my location.

I will break this down into topics or manipulations that are commonly discussed whether I do or do not do them.

Another thing that matters is the race or the breeding. Mine are all mutts, but they run from brown to black and are Northern bred survivor stock.

I'll break it down by items and actions:

## Mouse Guards

Typical questions are what to use and when to use them. I have only upper entrances so mouse guards are not an issue anymore. Back when I had lower entrances I used $1/4$" hardware cloth for mouse guards, but I might consider, if I were still using lower entrances, a popular device here in Southeast Nebraska. The device is a 3" to 4" wide piece of $3/8$" plywood cut to fit the width of the entrance and three $3/8$" laths cut to the 3" or 4" width of the plywood. This slides into the entrance reducing it to $3/8$" and forming a baffle so that the wind doesn't blow in. People who use it say there is no problem with mice as the $3/8$" gap being several inches long seems to deter the mice. They leave them on all year around.

As far as when, I'd try to get them on by or shortly after the first frost. Here we get some warm weather after the first frost, so the mice usually don't move in until it stays cold for several days. You want them on before then or the mice may already be in the hive. The other nice thing about the "baffle" type of entrance reducer/mouse guard is you can leave it in all year around and you don't have to worry about remembering to get the mouse guards on.

## Queen Excluders

I don't use excluders, but when I did, I would remove them before winter as they can cause the queen to get stuck below the excluder when the bees move up. The excluder will not stop the bees from moving up, but will keep the queen from joining them. You can store it on top of the inner cover or at the top of the hive I you like, but don't leave it between any boxes.

## Screened bottom boards (SBB)

I have these on about half of my hives. If the stand is short enough and enough grass blocks the wind, I sometimes leave out the tray, but usually I put the tray in. Some people in some climates seem to think it's good to leave them open year round, but I don't think it works well in a cold windy climate like mine. I also don't think the SBB helps much with Varroa, but it does help with ventilation in the summer and it keeps the bottom board dry in the winter. On the other hand a solid bottom board can double as a feeder and a cover.

## *Wrapping*

I don't. I tried it once, but it seemed to seal in all the moisture and cause the boxes to remain soaking wet all winter, so I quit doing it. If I were to try it again, which I probably won't, I'd put some wood on the corners to create an air space between the wood and the wrap.

## *Clustering hives together*

I put my hives on stands that hold two rows of seven (eight frame) hives. Basically they are eight foot long treated two by fours with four foot ends on them so the entire stand is 99" long (8' 3") because of the end pieces. The rails (the eight foot long pieces) are such that the outside ones are 20" from the center and the inside ones are 20" from the outside. This allows the hives (which are $19^7/_8$") to be all the way forward in the summer to maximize convenience of manipulating them, and all the way back in winter to minimize exposed area. So during the winter 10 of the hives are touching on three sides and the four on the outside ends are touching on two sides. This minimizes exposed walls. Sort of like huddling together for warmth.

## *Feeding Bees*

Contrary to popular belief, winter feeding honey or syrup does not work in Northern climates. Once the syrup doesn't make it above 50º F during the day (and it takes a while to warm up after a chilly night) the bees won't take it anymore anyway. The time to feed if needed is September, if necessary and if you're lucky you may be able to continue into October some years.

The questions always seem to be what concentration and how much.

When feeding honey, I don't water it down at all. Watered down it spoils quickly and I can't see wasting honey. When feeding syrup (because you have no honey or don't want to feed what you already went to the work of harvesting) the concentration should not be below 5:3 nor above 2:1. Thicker is better as it will require less evaporation, but I have trouble getting 2:1 to dissolve.

"How much" is not the right question. The right question is "what is the target weight?" For a large cluster in four medium eight frame boxes (or two ten frame deep boxes) should be between 100 and 150 pounds. In other words if the hive weighs 100 pounds, I might or might not feed, but if it weighs 150 I won't. If it weighs 75 pounds I'll try to feed 75 pounds of honey or syrup. Once the target weight is reached I would stop.

My management plan is to leave them enough honey and steal capped honey from other hives if they are light. But some years when the fall flow fails, I have to feed. I like to wait until the weather turns cold before harvesting as it solves several issues. 1) no wax moths to worry about. 2) the bees are clustered below so no bees to remove from supers. 3) I can assess better what to leave and what to take as the fall flow did or did not occur. Another option for a light hive, if it's not too light, is to feed dry sugar. The down side is that sugar is not stored like syrup, so it's more of an emergency ration, but the up side is you don't have to make syrup, buy feeders, etc. But it not being stored is also the up side. If they don't need it, you don't have syrup stored in your combs. You just put an empty box on the hive with some newspaper on the top bars and pour the sugar on top of the newspaper. I wet it a bit to clump it and wet the edge to get them to see it is food. If the hive is only a little light this is nice insurance. But if it's very light, I think they need to have some capped stores and I'd feed them honey or syrup.

A solid bottom board can be converted to a feeder. This makes sense to me because feeding isn't my normal management plan, leaving honey is. Why buy feeders for all your hives if feeding isn't a normal situation? This is not the best feeder, but it is the cheapest (basically free). If I need to feed, I don't have to buy a feeder for each hive. They hold about as much as a frame feeder.

Around here candy boards are popular, but the dry sugar on top is easier as you don't have to make the boards, and make the candy. You just use your standard boxes and sugar. I've also been known to spray syrup into drawn comb to give a light hive to get them through.

**Insulation**

Sometimes I insulate the tops and sometimes I don't. I gave up insulating anything else. I think it's a good idea to insulate the top, but I just don't always get it done. Since I run a simple top with a top entrance, when I do insulation it's just a piece of Styrofoam on top of the cover with the brick on top of that. This will reduce condensation on the top, as does the top entrance. Any thickness of Styrofoam will do. The main issue is condensation on the lid. When I have tried insulating the entire hive the moisture between the insulation and the hive became a problem.

## Top Entrances

I think this is essential to reducing condensation in my climate. It was not necessary when I was in Western Nebraska which is a much drier climate. It doesn't have to be a large top entrance, just a small one will do. The notch that comes on the notched inner covers is fine. This also provides a way for the bees to exit for cleansing flights on warm snowy days when the bottom entrance (which I don't have) would be blocked with snow. I have only top entrances and no bottom entrances.

## Where the cluster is

Usually around here it's in the top box going into and coming out of winter, with or without a top entrance. Sometimes it's not, but that seems to be the norm, despite what all the books seem to say. I leave them where they are and I don't try to make them be where I think they should be. Usually they spend the entire winter there. I would move them to one end in a horizontal hive, though, so they don't get to one end and starve with stores at the other end.

## How strong?

This question comes up a lot. I used to combine weak hives and I seldom lost a hive over winter. However, since I started trying to overwinter nucs I've realized how well a small hive takes off if it does make it through the winter. So I've overwintered much smaller clusters. Also if you have local queens, instead of southern queens, they do better as well as the darker bees overwintering on smaller clusters than the lighter colored bees. So, while I've never seen a softball sized cluster of southern package Italians get through the winter, I've seen that size of feral survivor stock, Carniolans and even Northern raised Italians make it. This is actually going into winter on a cold day (tight cluster). There is some attrition in the fall, and if they are this size in September and there is no flow and they are rearing no brood, they probably wouldn't make it. A strong Italian hive going into winter would be a basketball sized cluster or more, while Carniolans or Buckfasts are usually more like soccer ball sized or smaller, and feral survivors tend to be even smaller.

**Entrance reducers**

I do like them on all the hives. On the strong hives they create a traffic jam in the case of a robbing frenzy which will slow things down, and on a weak hive they create a smaller space to guard. On all the hives they create less of a draft than a wide open entrance. In fact when I have forgotten to open up the reducers in the spring, even the strong hives with the traffic jams because of it seem to do better than the ones that are wide open. I do try to remember to open them up on the strong hives for the main flow.

**Pollen**

I have, in recent years, started feeding pollen in the fall during a dearth so they are well stocked with pollen going into winter and so they have one more turnover of brood before winter sets in. There is no point in doing this while real pollen is coming in. I feed real pollen if I have enough. I have sometimes mixed it 50/50 with substitute or soybean flour when I'm desperate and don't have enough. I never mix it at less than 50% real pollen. You can trap this yourself or buy it from one of the suppliers like Brushy Mountain I feed it in the open. I put it on a SBB on top of a solid bottom board in an empty hive. This would be in September usually.

**Windbreak**

Some people use straw bales to get a windbreak. I hate mice and they seem to me to be mouse nests waiting to happen, so I don't. But if you kept them back a ways maybe they would work. I suppose one could use corn cribbing or snow fence for a wind break as well as any kind of privacy fence. Mel Disselkoen uses a ring of sheet metal around four hives to make a windbreak for them. This looks like a good setup to me but requires buying the metal and storing it during the rest of the year and then setting it up again in the fall.

**Eight frame boxes**

I find that eight frame boxes overwinter better than ten frame boxes. The width is more the size of a tree and the size of a cluster, so there is less food left behind. This is not to say that you can't winter bees in ten frame boxes, just that they seem to do slightly better in eight frame boxes.

**Medium boxes**

I find that medium boxes overwinter better than deeps as there is better communication between frames because of the gap between the boxes. If you picture what is in the hive when the bees cluster in the winter there are combs making walls between parts of the cluster. With a sudden cold snap a group of bees often get trapped on the other side of a deep frame when the cluster contracts as they can't get to the top or bottom and over, where with the medium the cluster usually spans the gap between the boxes providing communication between frames throughout the hive. Again, this is not to say you can't overwinter them in deeps, but only that they seem to do slightly better in mediums.

**Narrow frames**

I find they winter better on narrow frames ($1\frac{1}{4}$" on center instead of the standard $1\frac{3}{8}$" on center or the 9 frame arrangement in a ten frame box which is about $1\frac{1}{2}$" on center) because it takes less bees in the late winter to cover and keep the brood warm than it does with larger gaps. Again, this is not to say you can't overwinter them on $1\frac{3}{8}$" frames, only that they seem to do slightly better, build up earlier, get less chilled brood and less chalkbrood on narrow frames.

**Wintering Nucs**

I have tried overwintering nucs every winter since 2004. I can't claim to be good at it, but when I get nucs through they are my best hives the next year. I've tried many things from wrapping, huddling, heating, feeding syrup all winter etc. I've come to these conclusions. First, wrapping just made them too wet. Feeding syrup all winter did also. Insulating top and bottom and huddling were helpful. A heater if not too hot, down the middle of this arrangement was helpful, except every year someone unplugs it during the coldest spell, so it really hasn't helped. My nucs are a bit backwards of most as mine are combines of mating nucs rather than splits from my strong hives or requeening and splits from my weak hives. I've concluded that one mistake I've been making is I need to combine them soon enough for them to get reorganized as their own colony before the cold weather sets in. Which means about the end of July or the first of August. This also lets them get some stores put away and arranged the way they want. But assuming you're making splits of your weak hives and requeening them, the same rule holds true. You want them to have time to get organized as a colony. I'm liking the sugar on top more and more for these as feeding syrup has the problem of too much moisture. But if you feed early this isn't so much of a problem. Rather than spend a lot of time making special equipment for overwintering nucs, I think it's more practical to figure out how to overwinter them in your standard equipment. Granted, this makes more sense when your typical box is the size of a five frame deep nuc (my eight frame mediums are exactly that volume), but I hate having a lot of specialized equipment around when I can have equipment that is more multipurpose. My bottom board feeders work well for wintering nucs as you can stack up the nucs and see if they need to be fed and feed any of them without unstacking them.

**Banking queens**

I've tried overwintering a queen bank. I have not had really great success but these are the things that helped. You have to keep it warm enough to keep them from clustering or they will contract to the point that many of the queens will die. The best way I found to do this was a terrarium heater under the bank. You also have to repopulate the hive part way through the winter. This means either sacrificing one of the nucs or stealing some bees from a really strong hive. If you pull out a frame that is well covered in bees, but not too close to the center you have a better chance of *not* getting the queen and then you add that frame to the queen bank. If you get half of the queens through the winter, I think you're doing well. But if you do, you have a bunch of queens in the spring for queenless hives, splits and for selling at the time when the demand is high.

### Indoor wintering

I have not tried it other than the observation hive I typically winter. I have corresponded with many people who have tried it and it is far trickier than one would think. Bees need a cleansing flight now and then so they need to be free flying. They need temps down around 30º to 40º F to keep them inactive so they don't burn up all their stores and burn out from activity (inactive bees live longer than active bees). Ventilation and keeping bees cool enough seem to be the bigger issues with this than keeping them warm.

### Wintering observation hives

I have wintered an observation hive many times. The issues are to make sure they are strong enough going into winter. Have some way to feed them syrup. Have some way to feed them pollen. Don't' over feed the pollen. Make sure they are free flying (check the tube to make sure they haven't clogged it with dead bees and pollen). No, they won't all fly out and die because they are warm and confused about the weather outside. Some will no matter what, but that's just normal. They are quite aware of the weather outside. If they get too weak in the spring you may have to boost them with some bees. A handful or two of bees in an empty box that is connected to the tube will usually result in those bees moving into the hive without you having to take it outside and open it.

# Spring Management

**Tied to climate**

Next to wintering this seems to be the next biggest topic of discussion. And, next to wintering, this seems to be the most tied to climate. I can really only share with any confidence what I've actually experienced in my climate. Most places I've had bees are similar (cold winters etc.) but some were a bit colder (Laramie) and some a bit drier (Laramie, Brighton and Mitchell). But all in all most of my experience is in either the Panhandle of Nebraska or Southeast Nebraska. So keep that in mind.

**Feeding Bees**

Spring is a very volatile and unpredictable time here. We could have warm sunny flying weather and tree pollen as early as late February, but sometimes it stays cold until April. Our first actual nectar availability of any size, is the early fruit trees somewhere between early and late April, with mid April being most likely. The thing that seems to set off spring build up the most is pollen. Feeding syrup is iffy at best. If you feed syrup in February or March (if it every warms up enough to do so) and they decide to brood up a lot and we get a hard freeze (sub zero would not be unusual around here) then they could die from trying to keep the brood warm. On the other hand if they don't get going before the first nectar flow in mid April they won't build up enough to make a good crop. I like to just make sure they have pollen and stores. Dry sugar can stave off starvation. If the weather stays warm enough and they are light enough I might try syrup. I would still stick with 2:1 or 5:3 and not 1:1. 1:1 is just too much moisture in the hive and it doesn't keep well. So my main spring management up until the first blooms is to make sure they have pollen and they don't starve from lack of honey. Once the early flow starts, there is no need to feed really, but if it stays rainy for long periods it might pay. My bottom board feeders are easy enough to feed with on the fly like this. Just put in the plugs and fill with syrup even if it's raining. It helps to have a cover to keep the rain out of the syrup if it's really pouring, but if it's just drizzling, the 2:1 will work well and even if it gets watered down the bees still seem pretty interested as it gets diluted, all the way up to 1:2 or more.

## Swarm Control

The next issue in spring is heading off swarming. Of course you keep enough supers on that they don't run out of room. But in my experience, this alone will not head off swarming. You need some way to convince them that swarm preparation is not what is happening. If my bees had overhead honey, as Walt Wright's seem to in Tennessee, then I think I would do Checkerboarding/Nectar Management. But since mine are virtually always in the top box and I don't have capped honey to checkerboard above them, I just try to keep the brood nest open. In April, they are usually too small to swarm, but if they get going a lot, I'd put more boxes on. They only seem to swarm in April if they get overcrowded. In May is when I have to deal with swarm prevention in my location. The ideal is to keep them from swarming without splitting so you can have a maximum work force to make honey. In order to do this, I recommend keeping the brood nest open. Checkerboarding is fine for this, but as I say I don't seem to have the same conditions that lend well to this. So if a hive is getting really booming and strong from about early May on, I open up the brood nest. I do this with empty frames. No foundation. Just empty frames. Put these in the middle of the brood nest and they are quickly drawn and filled with brood. How many will depend on the strength of the hive. But if the nights aren't that chilly anymore and they can easily fill the gap where I intend to put the empty frame with festooning bees, then I can put another in. The maximum, which should only be done on a really strong hive, is an empty frame every other frame. The minimum, other than none, is one frame.

For more information on swarm prevention see the chapter *Swarm Control*.

## Splits

If you want more bees and honey isn't your prime consideration then do splits. Sometime on some warm days in April I will try to get all the way to the bottom board and clean it off while looking through the hive for brood, eggs, etc. to make sure things are going well. Other than that I just judge the strength and rate at which the population is increasing. Until you get good at judging this at a glance, look for swarm cells. Usually you can tip a box up and find them hanging down from the bottom of the frames. In the long run, this will give you an idea how much critical mass causes them to swarm and you can judge better how much to intervene. If you have swarm cells though, you already missed the opportunity for a large crop and now you need to worry about making splits.

## Supering

Of course you need to add supers. You don't want to do this when the hive is still struggling and the weather is cold, but once they are building up you need to add them. Doubling the space of the hive is my goal. If they are two boxes full, then I add two boxes. If they are four boxes full, then I add four boxes. Of course you eventually may, in a bumper crop year, get so tall you can't do this anymore, but it's a good way to try not to run out of room without giving them more room than they can handle.

# Laying Workers

*Cause*

When the hive is queenless, and therefore broodless, for several weeks sometimes some workers develop the ability to lay eggs. It's not actually the lack of a queen, but the lack of brood. But the lack of brood is caused by the lack of a queen. These are usually haploid (infertile with a half set of chromosomes) and will all develop into drones.

*Symptoms*

Laying workers lay eggs in worker cells, in addition to drone cells and usually lays several in each cell. Laying worker eggs are usually on the side of the cell instead of the bottom except in drone cells. A hive with lots of drones is a symptom of laying workers as are the multiple eggs in the cell and eggs on top of pollen.

Sometimes a queen, when she starts laying after a time of not laying, will lay a few double eggs but she usually stops after a day or two. The laying workers will lay three or four or more to a cell in almost every cell. The difficulty is that the bees think they have a queen (the laying workers) and will not accept one. The laying workers are virtually impossible to find. I have found one in a two frame nuc by studying every bee until I saw one lay, but this is impractical in a full sized hive since there would be too many bees and too many laying workers.

## Solutions

### Simplest, least trips to the beeyard

### Shakeout and forget

In my opinion there are only two practical solutions. The simplest solution if you have several hives and especially if the laying worker hive is a long trip, is just shake all the bees in front of the various other hives and divvy all the combs out to the other hives. This is my preferred method for an outyard or a small hive. It doesn't waste your time and money trying to requeen a hive that is going to reject the queen anyway. This is the method of least time spent on interventions and most predictable outcome.

If you really want to have that many hives, you can pull some frames from them several weeks after the shake out and do a split with some brood from all or several of your hives. A frame of open brood and emerging brood and honey and pollen from each and you'll have a nice split.

### Most successful but more trips to the beeyard

### Give them open brood

The only other really practical method, in my opinion, is to add a frame of open brood every week until they rear a queen. Usually by the second or third frame of open brood they will start queen cells. This is simple enough when the hive is in your backyard. Not so easy in an outyard 60 miles away.

## Other less successful or more tedious methods

I would do one of the above, but if you want to know every possible method that I've tried, here are the things I have done that sometimes work. Note some appear to be, and are, slight variations of the same theme.

1) If you have several weak laying worker hives and at least one strong queenright hive, put all the laying worker hives on the strong queenright hive. The resulting confusion between several hives will usually settle down to one queenright hive.

2) Any setup where a queen right hive is on the other side of a double screen so the brood pheromones get to the laying workers for two or three weeks will work to suppress them and then any introduction method will work get them queenright.

3) Put a queen cell in (either a frame from a hive trying to supersede or swarm or one that you made by queen rearing techniques). Sometimes they will let the queen emerge. Sometimes they will tear it down.

4) Put a virgin queen in. Just smoke it heavily and run her in. Sometimes they will accept her. Sometimes they will ball her.

5) Put a frame of emerging brood with a queen in a push in cage in the laying worker hive. When they are no longer biting the cage and killing the emerging attendants, release her. This usually works. Sometimes they will kill the queen.

### *More info on laying workers*

### Brood pheromones

It's the pheromones from open brood that suppress the laying workers from developing, but some do anyway. It is *not* the queen pheromone as many of the older books suggest.

See page 11 of Wisdom of the Hive:

> *"the queen's pheromones are neither necessary nor sufficient for inhibiting worker's ovaries. Instead, they strongly inhibit the workers from rearing additional queens. It is now clear that the pheromones that provide the proximate stimulus for workers to refrain from laying eggs come mainly from the brood, not from the queen (reviewed in Seeling 1985; see also Willis, Winston, and Slessor 1990)."*

There are always multiple laying workers even in a queenright hive

"Anarchistic bees" are ever present but usually in small enough numbers to not cause a problem and are simply policed by the workers *unless* they need drones. The number is always small as long as ovary development is suppressed.

See page 9 of "The Wisdom of the Hive"

> *"All studies to date report far fewer than 1 % of workers have ovaries developed sufficiently to lay eggs (reviewed in Ratnieks 1993; see also Visscher 1995a). For example, Ratnieks dissected 10,634 worker bees from 21 colonies and found that only 7 had moderately developed egg (half the size of a completed egg) and that just one had a fully developed egg in her body."*

If you do the math, in a normal booming queenright hive of 100,000 bees that's 70 laying workers. In a laying worker hive it's much higher.

# More than Bees

A honey bee colony is more than just bees. There is a whole ecology from microscopic to fairly large there are many symbiotic plus some benign relationships in the ecology of a bee colony. Even those benign relationships often crowd out pathogenic organisms.

**Macro and Microfauna**

For instance, there are over 32 kinds of mites that live in harmony with bees. When these are allowed to live (instead of being killed by acaracides) there are insects in the hive that eat them, such as the pseudoscorpion which also eat the malignant mites.

An examination of feral colonies shows just in the macroscopic arena the colony is full of life forms as diverse as mites, beetles, waxworms, ants and roaches.

**Microflora**

There are many microflora that live in the bees and in the colony. These vary from fungi to bacteria to yeasts. Many are necessary for the digestion of pollen or the maintenance of a healthy digestive tract by crowding out pathogens that would otherwise take over. Even seemingly benign ones and sometimes even mildly pathogenic ones often serve a beneficial purpose by supplanting otherwise deadly ones.

Many of the Lactobacillus genus are needed to properly digest the pollen and many of the Bifidobacterium and Gluconacetobacter genus are beneficial in the sense that they crowd out Nosema and other pathogens and probably contribute to digestion as well.

### Pathogens?

Even some seemingly pathogenic organisms such as Aspergillus fumigatus, which causes stonebrood, supplants worse pathogens, in this case Nosema. Or Ascosphaera apis, which causes chalkbrood but prevents European foulbrood.

### Upsetting the Balance

How much do we upset the balance of this rich ecosystem when applying anti-bacterials such as tylan or terramycin and anti-fungals such as Fumidil? Even essential oils and organic acids have anti-bacterial and anti-fungal effects. Then we kill off many of the mites and insects with acaracides.

After totally unbalancing this complex society of diverse organisms with no regard for benefit or not and contaminating the wax that we reuse and put in the hives as foundation, we are surprised to find that the bees are failing. Under such circumstances I would be surprised to find them flourishing!

### For More Reading

Try an internet search on the following phrases and read some of what comes back:

 bees microflora (10,900 hits)
 bees "symbiotic mites" (30 hits)
 bees symbiotic bacteria (25,100 hits)

Here are a few of the specific strains and groups you might want to research further:
 Bifidobacterium animalis
 Bifidobacterium asteroides

Bifidobacterium coryneforme
Bifidobacterium cuniculi
Bifidobacterium globosum
Lactobacillus plantarum
Bartonella sp.
Gluconacetobacter sp.
Simonsiella sp.

# Bee Math

All of the numbers about the life cycle of bees may seem irrelevant, so let's put them in a chart here and talk about what they are useful for.

| Caste | Days Hatch | Cap | Emerge | |
|---|---|---|---|---|
| Queen | $3^1/_2$ | 8 +-1 | 16 +-1 | Laying 28 +-5 |
| Worker | $3^1/_2$ | 9 +-1 | 20 +-1 | Foraging 42 +-7 |
| Drone | $3^1/_2$ | 10 +-1 | 24 +-1 | Flying to DCA 38 +-5 |

If you find eggs, and no queen how long ago do you *know* there was a queen? At least there was one three days ago and possibly is one now.

If you find just hatched larvae and open brood but no eggs when was there a queen? Four days ago.

If you put an excluder between two boxes and come back in four days and find eggs in one and not the other, what do you know? That the queen is in the one with eggs.

If you find a capped queen cell, how long before it should have emerged for sure? 9 days, but probably eight.

If you find a capped queen cell, how long before you should see eggs from that queen? Probably 20 days. Possibly as much as 29.

If you killed or lost a queen, how long before you'll have a laying queen again? Probably 24 days.

If you start from larvae and graft, how long before you need to transfer the larvae to a mating nuc? 10 days. (day 14 from when it was laid)

If you confine the queen to get the larvae how long before you graft? Four days.

If you confined the queen to get the larvae how long before we have a laying queen? 28 days.

# Races of Bees

## *Italian*

Apis mellifera ligustica. This is the most popular bee in North American. These, as all of the commercial bees, are gentle and good producers. They use less propolize than some of the darker bees. They usually have bands on their abdomen of brown to yellow color. Their biggest weakness is that they are prone to rob and drift. Most of these (as all of the queens) are bred and raised in the south, but you can find some northern breeders.

## *Starline*

These are just hybrid Italians. Two strains of Italians are kept separate and their hybrid is what the Starline queen is. They are very prolific and productive, but subsequent queens (supersedures, emergency and swarms) are disappointing. If you buy Starlines every year to requeen they will give you very good service. Unfortunately I don't know of any available anymore. They used to come from York and before that Dadant.

## *Cordovan*

These are a subset of the Italians. In theory you could have a Cordovan in any breed, since it's technically just a color, but the ones for sale in North American that I've seen are all Italians. They are slightly more gentle, slightly more likely to rob and quite striking to look at. They have no black on them and look very yellow at first sight. Looking closely you see that where the Italians normally have black legs and head, they have a purplish brown legs and head.

## Caucasian

Apis mellifera caucasica. They are silver gray to dark brown. They do propolis extensively. It is a sticky propolis rather than a hard propolis. The build up a little slower in the spring than the Italians. They are reputed to be more gentle than the Italians. Less prone to robbing. In theory they are less productive than Italians. I think on the average they are about the same productivity as the Italians, but since they rob less you get less of the really booming s that have robbed out all their neighbors.

## Carniolan

Apis mellifera carnica. These are darker brown to black. They fly in slightly cooler weather and in theory are better in northern climates. They are reputed by some to be less productive than Italians, but I have not had that experience. The ones I have had were very productive and very frugal for the winter. They winter in small clusters and shut down brood rearing when there are dearths.

## Midnite

These are, sort of, to the Caucasians what the Starline is to the Italians. At first they were two lines of Caucasians that were used to make an F1 cross. Later when the lines were hard to maintain, they were a Carniolan line crossed with a Caucasian line. They have that hybrid vigor that disappears in the next generation of queen. York used to sell them and before them Dadant. I don't know where they are available anymore.

### Russian

Apis mellifera acervorum or carpatica or caucasica or carnica. Some even say they are crossed with Apis ceranae (very doubtful). They came from the Primorksy region of Russia. They were used for breeding mite resistance because they were already surviving the mites. They are a bit defensive, but in odd ways. They tend to head butt a lot while not necessarily stinging any more. Any first cross of any race may be vicious and these are no exception. They are watchful guards, but not usually "runny" (tending to run around on the comb where you can't find the queen or work with them well). Swarminess and productivity are a bit more unpredictable. Traits are not well fixed. Frugality is similar to the Carniolans. They were brought to the USA by the USDA in June of 1997, studied on an island in Louisiana and then field testing in other states in 1999. They went on sale to the general public in 2000.

### Buckfast

These are a mixture of bees developed by Brother Adam of Buckfast Abbey. I had them for years. They were gentle. They built up rapidly in the spring, produced awesome crops and dropped back in population in the fall. They are just like the Italians as far as robbing. They are resistant to the tracheal mites. They are more frugal than the Italians, but not as much as the Carniolans.

### German or English native bees

Apis mellifera mellifera. These are the bees native to England or Germany. They have some of the characteristics of the other dark bees. They do well in damp cold climates. They tend toward being runny (excitable on the combs) and a bit swarmy, but also seem to be well adapted to Northern climates. Some of the ones that were here in the US were very unmanageable as far as temperament possibly because of crosses with the Italians.

### LUS

Small black bees similar to Carniolans or Italians in production and temperament but have mite resistance and have the ability of a laying worker to raise a new queen. This ability is called thelytoky. Several studies were done on them by the USDA in the 80's and 90's.

### Africanized Honey Bees (AHB)

I have heard these called Apis mellifera scutelata But Scutelata are actually African bees from the Cape. Dr. Kerr, who bred them thought they were Adansonii. AHB are a mixture of African (Scutelata) and Italian bees. They were created in an attempt to increase production of bees. The USDA bred these at Baton Rouge from stock obtained from Kerr from July 1942 until 1961. From the records I've seen it looks like the USDA shipped these queens to the continental US at about 1,500 queens a year from July 1949 until July of 1961. The Brazilians also were experimenting with them and the migration of those bees has been followed in the news for some time. They are extremely productive bees that are extremely defensive. If you have a hot enough that you think they are AHB you need to requeen them. Having angry bees where they might hurt people is irresponsible. You should try to requeen them so no one (including you) gets hurt.

# Moving Bees

**Moving hives two feet**

If you want to move a hive two feet, just stack the boxes off onto some kind of board (top, bottom etc.) and restack on the new location. Stacking them off and then restacking is so they are in the right order.

**Moving hives two miles**

If you want to move a hive two miles or more, you need to anchor the hive together for the trip and you need to load it. Since I am usually doing this by myself I will give instructions from that view.

I do this when the bees are flying. First I put my transportation as close as I can get to the hive. Directly behind it is best. I have a small trailer I often use, but a pickup would work too. I put a bottom board in the trailer where I think I want the hive to be. I put a strap under it so I can strap the hive together. You can buy small ones at the hardware store but they also sell them at bee supply places. I stack the boxes on the bottom board as I take them off. This leaves the hive in reverse order which will get reversed back when we unload. After all the boxes are on you need to nail all of the boxes together somehow. They sell 2" wide staples that can be used, or you can cut small ($2^1/_2$") squares of plywood and nail it between the parts of the hive to attach it all together. Cut a piece of #8 hardware cloth the length of the entrance and fold it into a 90 degree. It should fit tight enough to keep the bees in. Leave the entrance open until you are ready to leave.

Strap it together tightly and tie it anyway you need to or wedge it with empty bee boxes so that the hive can't shift or tip over on a curve or a sudden stop.

Next, you need to take into account your situation. If you have other hives at this location and the hive you are moving could lose a few foragers without hurting it much, just close it up and go. The returning foragers will find another hive. If this is your only hive or you are really concerned about losing foragers, then wait for dark and then close it up and go.

When you get to the new location, if it's already daylight, just unload the hive by putting a bottom board on the new location, removing the staples or plywood and stacking the boxes off onto it. If it's dark, wait for daylight and do the same thing.

Put a branch in front of the entrance so any bee leaving notices it. A green sapling with some leaves is nice so they have to fly through the middle of it. It causes them to stop and pay attention and reorient. This is useful at any distance of moving.

Other variations on this are a board (as mentioned in Dadant's The Hive and the Honey Bee) or grass plugging the entrance as mentioned many places.

> *"Bees moved less than a mile are likely to return in considerable numbers to their old location. This can be minimized by throwing grass or straw over their entrances to force them to take note of the change when they emerge for the first time from the hive at its new location"* —The How-To-Do-It book of Beekeeping, Richard Taylor

### *More than 2 feet and less than 2 miles*

This is a subject apparently full of controversy. There is an old saying that you move a hive 2 feet or two miles. I often need to move them 100 yards more or less. I've never seen that it was a problem. I move hives as seldom as I need to because anytime you move a hive even two feet, it disrupts the hive for a day. But if I need to, I move them. I didn't invent all of the concepts here, but some of them I refined for my uses. Here is my technique.

It occurs to me that a lot of detail that is intuitively obvious to me may not be to a newbie. So here is a detailed description of how I usually move hives single handedly. This is assuming the hive is too heavy to move in one piece or I lack the help to do so. But it works so well, I don't even think about using other methods. But if you have help and can lift it, you can block the entrance and move it all at once at night and put a branch in front. I know every time I tell any version of this method, someone quotes the "2 feet or 2 miles" rule and says you can't do it and you can only move them two feet or you'll lose all your bees. I've done this many times with no noticeable loss of workforce and no bees clustering at the old location by the second night.

### *Moving hives 100 yards or less by yourself.*

## Concepts

### Reorientation

When bees fly out of the hive, normally, they pay no attention to where they are. They know where they live and don't even consider it on the way out. When they fly back they look for familiar landmarks and follow them home. They orient when they first leave the hive as a young bee, but only certain conditions cause them to reorient after that. One is confinement. Any confinement will cause some. 72 hours causes about the maximum reorientation. Any length of time more than that is difficult to tell any difference. A blockage of the exit causes reorientation. People sometime stuff the entrance with grass. This combines the act of removing it, which sets off reorientation, with some confinement, which causes some reorientation. An obvious obstruction that causes them to deviate from their normal exit will set off reorientation. A branch or a board in front of the entrance that causes them to have to fly around it, will cause them to pay attention to where they are. Some oldtimers would just bang the hive around really well to indicate to the bees that something has happened and they need to pay attention.

### Autopilot

When a bee is returning to the hive they tend to be on "autopilot". It's like you driving home from work. You don't think about where the turns are, you just make them. If they have done no reorientation, they will see landmarks and return to the old hive and have no idea where to go. If they have reoriented, they will still fly back to the old location, but when they see the hive isn't there, they think back to when they left and remember.

### Finding the new hive

Assuming they did not reorient and they have to figure out where the new hive is, then they have to do increasing spirals out until they smell the hive. Odds are they will move into the first hive they find doing this. How long it takes to find the new location is exponential to the distance. In other words if it's twice as far away it will take them four times as long to find it.

### Weather

Keep in mind that cold weather can complicate things in odd contradictory ways. On the one hand if they have been confined for 72 hours and you move them, they are most likely to reorient. On the other hand if they fly back to the old location they have to find the hive again before they get too chilled or they will die.

### Leaving a box

Leaving a box at the old location is another of those complicated things. If you leave one from the start they all return and just stay there. If you leave nothing at the old location they will look for the new location, but some may get stuck at the old location. If you wait until just before dark to put a box there you will motivate them to find the new location, but still give them somewhere to go. You can move that to the new location, and in warm weather, just set it beside the hive. In cold weather you may need to put this box on top, but that's not a pleasant thing to do in the dark.

## Materials:

- Second bottom board. If you don't have one, some board big enough to set the hives on will do.
- Third bottom board.
- A cover cloth is useful but not necessary. If you don't have one, some board big enough to set the hives on will do.
- Second lid. If you don't have one, any board big enough to put on top of the hive will work.
- Smoker.

- Veil.
- Gloves (optional but nice)
- Bee Suit (optional but nice)
- A branch that will stick up nicely and disrupt the flight of the bees leaving the hive.

**Method**

Suit up to your comfort level. Remember we will not be manipulating frames so the gloves are not a big disadvantage.

I usually put a puff of smoke in the entrance, then pull off the lid and put a puff in the inner cover (unless you don't have an inner cover).

Then I put four or five good strong puffs of smoke in the entrance and wait a minute. Then repeat four or five puffs and wait a minute. I do this until I see just a whiff of smoke out the top. This is more smoke than I usually use, but we will be rearranging this hive twice and I need it calm all the way through. If they are getting irate or you are moving an exceptionally strong and large hive and it is taking some time, feel free to smoke some more from time to time.

Wait about three minutes before opening the hive.

Set the second bottom board next to the hive. Take the top box off, lid and all and put it on the bottom board. Remove the lid and move each box from the old location to the new bottom board until you reach the last box. You don't need to restack the last one because we are moving it first. You now have reversed the order of the boxes so when we move them to the new location they will be in the correct order.

Put the second lid on the stack of boxes to keep the bees calm and the lid on the last brood box so they won't fly in your face. Carry the last brood box, with the lid and bottom board to the new location.

Put the branch in front of the entrance so that the bees have to fly through the branch. It doesn't have to be so thick they have trouble getting through it, just enough that they can't miss seeing it. This is to cause them to reorient when they leave. If you watch them they will start by circling the hive, then make larger circles until they have placed the hive in their mental map of their world. Since you have moved the hive to a new place and that place is within their known world they do this fairly quickly.

Remove the lid, if you want to use a cloth cover, put it on the brood box. It will help keep the bees calm, but you have to get it off with a box in your hands when you come back. That is why I like a cloth instead of a cover. Take the lid back to the old location. Take the top box and lid off and put in on the third bottom board. Put the lid you that you brought back on the stack of boxes. Again this is so there is always a lid on the stack of boxes and a lid on the box you are moving. This helps keep the bees calm. You may be thinking, that the bottom is exposed while you're carrying it. Yes, but the bees don't move down when they are getting jostled, they move up. Not that I'd wear shorts while moving the boxes.

Carry the second box over to the new location and catch the cloth (if you used one) with one finger while still holding the box and lift the cloth off and set the box down. Remove the lid and replace it with the cloth.

Go back to the old location with the lid and repeat until all of the boxes are at the new location.

We want nothing at the old location that looks like home. When it's almost dark we will take that last box back, to the old location with its own lid and bottom so you can bring it back to the old location just after dark.

After dark, block the entrance, or pull out the stick and carry it to the new location with the bottom in place. Just set it beside the hive with the branches in front of its entrance. Open the entrance or replace the stick. *Do not try to put this box on the hive in the dark unless the weather is cold!* If you have never opened a hive in the dark, consider yourself wise or fortunate and don't. The bees are *very* defensive after dark and will attack and cling and crawl on you looking for a way to sting.

The next morning you can put the last box on top of the hive. Remove any equipment from the old site so they don't start clustering there.

Some field bees will return to the old location. If they paid attention and reoriented, they will then remember where they hive was and go back to that new location. If not, they will circle until they find the new location and then will be fine after that.

You can check in the evening before dark and see if any are clustering at the old location. If so, put a super there and they will move into it and you can move them after dark again. I have never had any clustered there by the next day and seldom had any at all.

# Treatments for Varroa not working

A lot of you use some treatment, and your mite drops don't change much and you assume you're not killing mites. So let's just look at some numbers.

Independent of *what* the treatment is, here is just a rough idea of what goes on. These are round numbers and probably underestimate the mites' reproduction and underestimate how many get groomed off by the bees.

Assuming treating every week and a treatment with 100% effectiveness on phoretic mites. If you assume that half the Varroa are in the cells and you have a total mite population of 32,000, and if we assume half the phoretic mites will go back in the cells and in one week, half of the mites in the cells will have one offspring each and emerge then the numbers look like this:

| 100% Week | Phoretic | Capped | Killed | Reproduced | Emerged | Returned |
|---|---|---|---|---|---|---|
| 1 | 16,000 | 16,000 | 16,000 | 8,000 | 16,000* | 8,000 |
| 2 | 8,000 | 16,000 | 8,000 | 8,000 | 16,000 | 8,000 |
| 3 | 8,000 | 16,000 | 8,000 | 8,000 | 16,000 | 8,000 |
| 4 | 8,000 | 16,000 | 8,000 | 8,000 | 16,000 | 8,000 |

\* half of the 16,000 plus 8,000 offspring
Capped is inside capped cells. Returned is the number that went back into cells and got capped.

Now let's assume treating every week and 50% effectiveness on phoretic mites with all the other assumptions the same:

| 50% Week | Phoretic | Capped | Killed | Reproduced | Emerged | Returned |
|---|---|---|---|---|---|---|
| 1 | 16,000 | 16,000 | 8,000 | 8,000 | 16,000 | 12,000 |
| 2 | 12,000 | 20,000 | 6,000 | 10,000 | 20,000 | 13,000 |
| 3 | 13,000 | 23,000 | 6,500 | 11,500 | 23,000 | 14,750 |
| 4 | 14,750 | 26,250 | 7,375 | 13,125 | 26,250 | 16,813 |

Now let's assume treating once every week with 50% effectiveness with no brood in the hive:

| 50% Week | No Phoretic | Brood Capped | Killed | Reproduced | Emerged | Returned |
|---|---|---|---|---|---|---|
| 1 | 32,000 | N/A | 16,000 | N/A | N/A | N/A |
| 2 | 16,000 | N/A | 8,000 | N/A | N/A | N/A |
| 3 | 8,000 | N/A | 4,000 | N/A | N/A | N/A |
| 4 | 4,000 | N/A | 2,000 | N/A | N/A | N/A |

Then of course there's 100% with no brood:

| 100% Week | No Phoretic | Brood Capped | Killed | Reproduced | Emerged | Returned |
|---|---|---|---|---|---|---|
| 1 | 32,000 | N/A | 32,000 | N/A | N/A | N/A |
| 2 | N/A | N/A | N/A | N/A | N/A | N/A |
| 3 | N/A | N/A | N/A | N/A | N/A | N/A |
| 4 | N/A | N/A | N/A | N/A | N/A | N/A |

And no treatment would look like this:

| 0% Week | Phoretic | Capped | Killed | Reproduced | Emerged | Returned |
|---|---|---|---|---|---|---|
| 1 | 16,000 | 16,000 | N/A | 8,000 | 16,000 | 16,000 |
| 2 | 16,000 | 24,000 | N/A | 12,000 | 24,000 | 20,000 |
| 3 | 20,000 | 32,000 | N/A | 16,000 | 32,000 | 26,000 |
| 4 | 26,000 | 42,000 | N/A | 21,000 | 42,000 | 34,000 |

A real mathematical model, of course, should take into account a lot of things including drifting, robbing, hygienic behavior (chewing out), grooming, time of year etc. I was just hoping to get the general principle across of what is happening when you treat.

# A Few Good Queens

### Simple Queen Rearing for a Hobbyist

I get this question a lot, so let's simplify this as much as possible while maximizing the quality of the queens as much as possible.

### Labor and Resources

The quality of a queen is directly related to how well she is fed which is related to the labor force available to feed the larvae (density of the bees) and available food.

### Quality of Emergency Queens

First let's talk about emergency queens and quality. There has been much speculation over the years on this matter and after reading the opinions of many very experienced queen breeders on this subject I'm convinced that the prevailing theory that bees start with too old of a larvae is not true. I think to get good quality queens from emergency cells one simply needs to insure they can tear down the cell walls and that they have resources of food and labor to properly care for the queen. This means a good density of bees (for labor), frames of pollen and honey (for resources), and nectar or syrup coming in (to convince them they have resources to spare).

So if one adds either new drawn wax comb or wax foundation without wires or even empty frames to the brood nest during a time of year they are anxious to raise queens (from about a month after the first blooms until the end of the main flow), they quickly draw this comb and lay it full of eggs. So four to five days after adding it, there should be frames of larvae on newly drawn wax with no cocoons to interfere with them tearing down the cell walls to build queen cells. If one were to do this in a strong hive and at this point remove the queen on a frame of brood and a frame of honey and put it aside in a nuc, the bees will start a lot of queen cells.

**The experts on emergency queens:**

## Jay Smith:

> *"It has been stated by a number of beekeepers who should know better (including myself) that the bees are in such a hurry to rear a queen that they choose larvae too old for best results. Later observation has shown the fallacy of this statement and has convinced me that bees do the very best that can be done under existing circumstances.*
>
> *"The inferior queens caused by using the emergency method is because the bees cannot tear down the tough cells in the old combs lined with cocoons. The result is that the bees fill the worker cells with bee milk floating the larvae out the opening of the cells, then they build a little queen cell pointing downward. The larvae cannot eat the bee milk back in the bottom of the cells with the result that they*

*are not well fed. However, if the colony is strong in bees, are well fed and have new combs, they can rear the best of queens. And please note— they will never make such a blunder as choosing larvae too old."— Better Queens. Jay Smith*

## C.C. Miller:

*"If it were true, as formerly believed, that queenless bees are in such haste to rear a queen that they will select a larva too old for the purpose, then it would hardly do to wait even nine days. A queen is matured in fifteen days from the time the egg is laid, and is fed throughout her larval lifetime on the same food that is given to a worker-larva during the first three days of its larval existence. So a worker-larva more than three days old, or more than six days from the laying of the egg would be too old for a good queen. If, now, the bees should select a larva more than three days old, the queen would emerge in less than nine days. I think no one has ever known this to occur. Bees do not prefer too old larvae. As a matter of fact bees do not use such poor judgment as to select larvae too old when larvae sufficiently young are present, as I have proven by direct*

> *experiment and many observations."—Fifty Years Among the Bees,* C.C. Miller

## Equipment

Second let's talk about equipment. One can set up mating nucs in standard boxes with dummy boards (or division boards) but only if you have the extra boxes or division boards. The advantage is that you can expand this as the hive grows if you don't use the queen. You can also build either two frame boxes or divide larger boxes into two frame boxes (commonly sold as queen castles). These need to be the same depth as your brood frames.

## Method:

### Make sure they are well fed

Feed them for a few days before you start unless there is a strong flow on.

### Make them Queenless

So if we make a hive queenless (do what you like about having new comb or not) nine days after making them queenless these will be mostly mature and capped and be three days from emerging.

### Make up Mating Nucs

At this point unless you intend to use the cells to requeen your hives, we need to set up mating nucs. The "queen castles" or four way boxes that take your standard brood frames and make up four, two frame mating nucs in one box are very good for this, but dummy boards and regular boxes can work also. In my operation these are all medium depth two frame nucs. The queen we removed earlier goes well in one of these also. We now want a frame of brood and a frame of honey in each of the mating nucs.

## Transfer Queen Cells

The next day (ten days after making them queenless) we will cut out (with a sharp knife) the queen cells from the new wax combs we put in. If we used unwired foundation (or none) they should be easy to cut out without running into obstacles (as we would with wire and with plastic foundation) and can put each of the cells in a mating nuc. You can just press an indentation with your thumb and gently place the cell in the indentation. If you want you can also just put each frame that has cells on it in a mating nuc and sacrifice the extra cells (as the first queen out will destroy them). This is helpful if you have plastic foundation or you just don't want to mess with cutting out cells.

## Check for Eggs

Two weeks later we should see some eggs in the mating nucs. If not, then by three weeks we should. Let her lay up the nuc well before moving her to a hive or caging her and banking her for later.

Next round just make the mating nuc queenless again the day before adding cells.

Now that these nucs are well populated by the brood the queen has laid, we can make more queens by simply making a strong mating nuc queenless and they will raise more queens. Again, it's the density of bees and the supply of food that are the issues. We can also, if they are wax combs, cut cells out and make use of multiple cells in other mating nucs as well. In this case either set up those nucs the day before or remove the queen the day before.

And that is all there is to raising a few queens.

# About the Author

*"His writing is like his talks, with more content, detail, and depth than one would think possible with such few words...his website and PowerPoint presentations are the gold standard for diverse and common sense beekeeping practices."—Dean Stiglitz*

    Michael Bush is one of the leading proponents of treatment free beekeeping. He has had an eclectic set of careers from printing and graphic arts, to construction to computer programming and a few more in between. Currently he is working in computers. He has been keeping bees since the mid 70's, usually from two to seven hives up until the year 2000. Varroa forced more experimentation which required more hives and the number has grown steadily over the years from then. By 2008 it was about 200 hives. He is active on many of the Beekeeping forums with last count at about 50,000 posts between all of them. He has a web site on beekeeping at:
    www.bushfarms.com/bees.htm

# Index to *The Practical Beekeeper*

acaracide, 444, 445

Africanized honey bee, 387, 416
   cell size, 384
   Varroa mite, 387

Apis mellifera mellifera. *See* native bees

Apistan, 267

Baudoux, 361, 368

bee math, 447

beekeeping philosophy, 266

Brother Adam
   Buckfast bees, 451

cell size
   drone, 368
   Honey Super Cell, 367
   PermaComb, 367

chalkbrood
   *Ascosphaera apis*, 445

CheckMite, 267

comb, 369
   brood comb, 350, 353
   burr comb, 289
   cell size, 367
   drone comb, 290, 415
   messed up, 406
   natural comb, 371, 375, 385, 390, 393
   width (thickness), 372, 382

confine
   confine queen, 356

contamination, 393

coumaphos, 267, 407

drone, 352, 439
  cell size, 369, 371
  drone brood, 353, 410
  drone escape, 273
  frame spacing, 412, 415
  life cycle, 447
  when to split, 353

drone congregation area, 447

Dzierzon, Jan
  natural cell size, 382

eight frame medium, 279, 298, 432

feeder
  baggie feeder, 307
  Boardman feeder, 311
  bottom board feeder, 427, 432, 436
  considerations, 309
  frame feeder, 307, 310
  jar feeder, 311
  Jay Smith bottom board feeder, 316, 323
  Miller feeder, 314
  top feeder, 307
  types, 310–30

feeding
  honeycomb, 276
  honeycomb versus syrup, 285

feral bees, 378
  winter cluster, 419, 429

fluvalinate, 267, 407

foundation
  4.9 mm, 361
  5.4 mm, 361
  natural cell size, 361–85
  small cell, 361
  standard, 361

frame
  foundationless, 283, 394, 399, 408
  Mann Lake PF120, 364

narrow, 409–18
 spacing, 372, 384

hive
 clustering, 425
 ecology, 444–46
 horizontal. *See* horizontal hive
 moisture, 304, 425, 427, 436
 natural comb, 406
 nucleus hive, 431
 queenless, 439
 stores, 420
 top entrance, 272
 wintering, 423–34
 wrapping, 292, 425

hives
 painting, 294

horizontal hive, 280–81

Huber, François, 372, 375, 383
 natural cell size, 371
 natural frame spacing, 409

Hutchinson, W.Z., 261, 269, 304

inner cover
 top entrance, 428

Jay Smith
 queens, 465

Koover, Charles, 415

Langstroth, L.L., 400

laying worker, 439–43

Lusby, Dee and Ed, 367, 370, 371, 380, 388, 389, 391

mouse guard, 423
 top entrance, 273

moving bees, 454–61

narrow frames
 chalkbrood, 431

native bees, 451

natural cell size, 371

no treatment, 266, 371, 389, 463

Nosema
   microflora, 444

observation hive
   wintering, 433

pests
   adaptation to, 379
   resistance to treatment, 267

pollen trap
   drone access, 263
   top entrance, 333

propolis
   scraping, 293

queen
   emergency, 464–67

queen cells
   destroying, 349

queen excluder, 262
   winter, 420, 424

queen rearing, 464–69

races of honey bee, 449–53

requeening
   lazy beekeeping, 299

screened bottom board, 424

splits, 352–60
   cut down, 355
   cutdown/combine, 356
   even, 354
   swarm control, 354
   walk away, 354

starter strip, 377, 386, 403, 404

stimulative feeding, 302

stonebrood, 445

survivor stock, 388

swarm cell, 349, 421
    removing, 290
    split, 437

swarm control, 290, 345–51, 436
    cut down split, 349

switching hive bodies, 295

syrup, 330–31
    pH, 300
    winter feeding, 305, 306

Taylor, Richard
    burr comb, 288
    do nothing, 291
    lighter boxes, 279

thelytoky, 452

top bar hive, 282, 386

top entrance, 272–73, 333–40
    Elisha Gallup, 333
    preparing for winter, 428

uniform frame size, 274

Varroa mite
    breaking brood cycle, 356
    capping time, 375
    chemical treatment, 267
    natural cell size, 285, 371, 376, 393
    phoretic phase, 462
    seasonal management, 420–21
    treatment, 462

Wright, Walt, 351

yearly cycle, 419

CPSIA information can be obtained
at www.ICGtesting.com
Printed in the USA
BVHW081328100221
599764BV00001B/56